Teen guides to environmental science

P9-CLC-075

Chicago Public Library
C
P
L
REFERENCE

Teen Guides to

Environmental Science

Teen Guides to

Environmental Science

Resources and Energy

VOLUME II

John Mongillo

with assistance from Peter Mongillo

Greenwood Press
Westport, Connecticut • London

Library of Congress Cataloging-in-Publication Data

Mongillo, John F.
Teen guides to environmental science / John Mongillo with assistance from Peter Mongillo.
p. cm.
Includes bibliographical references and index.
Contents: v. 1. Earth systems and ecology—v. 2. Resources and energy—v. 3. People and their environments—v. 4. Human impact on the environment—v. 5. Creating a sustainable society.
ISBN 0-313-32183-3 (set : alk. paper)—ISBN 0-313-32184-1 (v. 1 : alk. paper)—
ISBN 0-313-32185-X (v. 2 : alk. paper)—ISBN 0-313-32186-8 (v. 3 : alk. paper)—
ISBN 0-313-32187-6 (v. 4 : alk. paper)—ISBN 0-313-32188-4 (v. 5 : alk. paper)
1. Environmental sciences. 2. Human ecology. 3. Nature–Effect of human beings on. I. Mongillo, Peter A. II. Title.
GE105.M66 2004
333.72—dc22 2004044869

British Library Cataloguing in Publication Data is available.

Library of Congress Catalog Card Number: 2004044869
ISBN: 0-313-32183-3 (set)
0-313-32184-1 (vol. I)
0-313-32185-X (vol. II)
0-313-32186-8 (vol. III)
0-313-32187-6 (vol. IV)
0-313-32188-4 (vol.V)

First published in 2004

Greenwood Press, 88 Post Road West, Westport, CT 06881
An imprint of Greenwood Publishing Group, Inc.
www.greenwood.com

Printed in the United States of America

The paper used in this book complies with the Permanent Paper Standard issued by the National Information Standards Organization (Z39.48–1984).

10 9 8 7 6 5 4 3 2 1

Contents

Acknowledgments

The authors wish to acknowledge and express the contribution of the many nongovernment organizations, corporations, colleges, and government agencies that provided assistance to the authors in the research for this book. The authors are grateful to the Greenwood Publishing Group for permission to excerpt text and photos from *Encyclopedia of Environmental Science*, John Mongillo and Linda Zierdt-Warshaw, and *Environmental Activists*, John Mongillo and Bibi Booth. Both books are excellent references for researching environmental topics and gathering information about environmental activists. Many thanks to those who provided special assistance in reviewing particular topics and offering comments and suggestions: Sara Jones, middle school director for La Jolla Country Day School in San Diego, California; Emily White, teacher of geography and world cultures at the 5th grade level at La Jolla Country Day School, San Diego, California; Lucinda Kramer and John Guido, middle school social studies coordinators, North Haven, Connecticut; Daniel Lanier, environmental professional, and Susan Santone, executive director of Creative Change, Ypsilanti, Michigan.

A special thank you goes to the following people and organizations that provided technical expertise and/or resources for photos and data: Neil Dahlstrom, John Deere & Company; Francine Murphy-Brillon, Slater Mill Historic Site; Lake Worth Public Library, Florida; Pacific Gas & Electric; Energetch; Environmental Justice Resource Center; NASA Johnson Space Center; Seattle Audubon Society; John Onuska, INMETCO; Cathrine Sneed, Garden Project; Denis Hayes, president, Bullitt Foundation; Ocean Robbins, Youth for Environmental Sanity; Maria Perez and Nevada Dove, Friends of McKinley; Juana Beatriz Gutiérrez, cofounder and president of Madres del Este de Los Angeles—Santa Isabel; Mikhail Davis, director, Brower Fund, Earth Island Institute; Randall Hayes, president, Rainforest Action Network; Tom Repine, West Virginia Geologic Survey; Peter Wright and Nancy Trautmann, Cornell University; Mary N. Harrison, University of Florida; and Huanmin Lu, University of Texas, El Paso.

Other sources include Centers for Disease Control and Prevention, Department of Environmental Management, Rhode Island; ChryslerDaimler; Pattonville High School; National Oceanic and

Atmospheric Administration; Chuck Meyers, Office of Surface Mining; U.S. Department of Agriculture; U.S. Fish and Wildlife Service; U.S. Department of Energy; U.S. Environmental Protection Agency; U.S. National Park Service; National Renewable Energy Laboratory; Tower Tech, Inc.; Earthday 2000; Marilyn Nemzer, Geothermal Education Office; U.S. Agricultural Research Service; U.S. Geological Survey; Glacier National Park; Monsanto; CREST Organization; Shirley Briggs, Vortec Corporation; National Interagency Fire Center/Bureau of Land Management; Susan Snyder, Marine Spill Response Corporation; Lisa Bousquet, Roger Williams Park Zoo, Rhode Island; Netzin Gerald Steklis, International National Response Corporation; U.S. Department of the Interior/Bureau of Reclamation; Bluestone Energy Services; OSG Ship Management, Inc.; and Sweetwater Technology.

In addition, the authors wish to thank Hollis Burkhart and Janet Heffernan for their copyediting and proofreading support; Muriel Cawthorn, Hollis Burkhart, and Liz Kincaid for their assistance in photo research; and illustrators Christine Murphy, Susan Stone, and Kurt Van Dexter.

The responsibility of the accuracy of the terms is solely that of the authors. If errors are noticed, please address them to the authors so that corrections can be made in future revisions.

INTRODUCTION

Teen Guides to Environmental Science is a reference tool which introduces environmental science topics to middle and high school students. The five-volume series presents environmental, social, and economic topics to assist the reader in developing an understanding of how human activity has changed and continues to change the face of the world around us.

Events affecting the environment are reported daily in magazines, newspapers, periodicals, newsletters, radio, and television, and on Websites. Each day there are environmental reports about collapsing fish stocks, massive wastes of natural resources and energy, soil erosion, deteriorating rangelands, loss of forests, and air and water pollution. At times, the degradation of the environment has led to issues of poverty, malnutrition, disease, and social and economic inequalities throughout the world. Human demands on the natural environment are placing more and more pressure on Earth's ecosystems and its natural resources.

The challenge in this century will be to reverse the exploitation of Earth's resources and to improve social and economic systems. Meeting these goals will require the participation and commitment of businesses, government agencies, nongovernment organizations, and individuals. The major task will be to begin a long-term environmental strategy that will ensure a more sustainable society.

CREATING A SUSTAINABLE SOCIETY

Sustainable development is a strategy that meets the needs of the present without compromising the ability of future generations to meet their own needs. Many experts believe that for too long, social, economic, and environmental issues were addressed separately without regard to each other. In creating a sustainable society, there needs to be an integration of goals related to economic growth, environmental protection, and social equity. Some of these integrated sustainable goals include the following:

- Improve the quality of human life
- Conserve Earth's diversity

- Minimize the depletion of nonrenewable resources
- Keep within Earth's carrying capacity
- Enable communities to care for their own environments
- Integrate the environment, economy, and human health into decision making
- Promote caretakers of Earth.

OVERVIEW

Teen Guides to Environmental Science provides an excellent opportunity for students to study and focus on the integration of ecological, economical, and social goals in creating a sustainable society. Within the five-volume series, students can research topics from a long list of contemporary environmental issues ranging from alternative fuels and acid rain to wetlands and zoos. Strategies and solutions to solve environmental issues are presented, too. Such topics include soil conservation programs, alternative energy sources, international laws to preserve wildlife, recycling and source reduction in the production of goods, and legislation to reduce air and water pollution, just to name a few.

Major Highlights

- Assists students in developing an understanding of their global environment and how the human population and its technologies have affected Earth and its ecology.
- Provides an interdisciplinary perspective that includes ecology, geography, biology, human culture, geology, physics, chemistry, history, and economics.
- "Raises a student's awareness of a strategy called sustainable development that meets the needs of the present without compromising the ability of future generations to meet their own needs" (Bruntland Commission). The strategy includes a level of economic development that can be sustained in the future while protecting and conserving natural resources with minimum damage to the environment. People concerned about sustainable development suggest that meeting the needs of the future depends on how well we balance social, economic, and environmental objectives—or needs—when making decisions today.
- Presents current environment, social, and economic issues and solutions for preserving wildlife species, rebuilding fish stocks, designing strategies to control sprawl and traffic congestion, and developing hydrogen fuel cells as a future energy source.

- Challenges everyone to become more active in their home, community, and school in addressing environmental problems and discussing strategies to solve them.

ORGANIZATION

Teen Guides to Environmental Science is divided into five volumes.

Earth Systems and Ecology

Volume I begins the discussion of Earth as a system and focuses on ecology—the foundation of environmental science. The major chapters examine ecosystems, populations, communities, and biomes.

Resources and Energy

Currently, fossil fuels drive the economy in much of the world. In Volume II conventional fuels such as petroleum, coal, and natural gas are reported. Other chapters elaborate on nuclear energy, hydrogen energy, wind energy, geothermal energy, solar energy, and natural resources such as soil and minerals, forests, water resources, and wildlife preserves.

People and Their Environments

The history of civilizations, human ecology, and how early and modern societies have interacted with the environment is presented in Volume III. The major chapters highlight the Agricultural Revolution, the Industrial Revolution, global populations, and economic and social systems.

Human Impact on the Environment

Volume IV discusses the causes and the harmful effects of air and water pollution and sustainable solution strategies to control the problems. Other chapters examine the human impact on natural resources and wildlife and discuss efforts to preserve them.

Creating a Sustainable Society

Volume V focuses on the importance of living in a sustainable society in which generations after generations do not deplete the natural resources or produce excessive pollutants. The chapters present an overview of sustainability in producing products, preserving wildlife habitats, developing sustainable communities and transportation systems, and encouraging sustainable management practices in agriculture and commercial fishing. The last chapter in this volume considers the importance of individual activism in identifying and solving environmental problems in one's community.

PROGRAM RESEARCH

The five-volume series represents research from a variety of recurring and up-to-date sources, including newspapers, middle school and high school textbooks, trade books, television reports, professional journals, national and international government organizations, nonprofit organizations, private companies, businesses, and individual contacts.

CONTENT STANDARDS

The series provides a close alignment with the fundamental principles developed and reported in the President's Council on Sustainable Development and the learning outcomes for middle school education standards found in the North American Association for Environmental Education, the National Geography Standards, and the National Science Education Standards.

MAJOR ENVIRONMENTAL TOPICS

The *Teen Guides to Environmental Science* provide terms, topics, and subjects covered in most middle school and high schools environmental science courses. These major topics of environmental science include, but are not limited to:

- Agriculture, crop production, and pest control
- Atmosphere and air pollution
- Ecological economies
- Ecology and ecosystems
- Endangered and threatened wildlife species
- Energy and mineral resources
- Environmental laws, regulations, and ethics
- Oceans and wetlands
- Nonhazardous and hazardous wastes
- Water resources and pollution.

SPECIAL FEATURES

Tables, Figures, and Maps

Hundreds of photos, tables, maps, and figures are ideal visual learning strategies used to enhance the text and provide additional information to the reader.

Vocabulary

The vocabulary list at the end of each chapter provides a definition for a term used within the chapter with which a reader might be unfamiliar.

Marginal Topics

Each chapter contains marginal features which supplement and enrich the main topic covered in the chapter.

Activities

More than 100 suggested student research activities appear at the ends of the chapters in the books.

In-Text References

Many of the chapters have specially marked callouts within the text which refer the reader to other books in the series for additional information. For example, fossil fuels are discussed in Volume V; however, an in-text reference refers the reader to Volume II for more information about the topic.

Websites

A listing of Websites of government and nongovernment organizations is available at the end of each chapter allowing students to research topics on the Internet.

Bibliography

Book titles and articles relating to the subject area of each chapter are presented at the end of each chapter for additional research opportunities.

Appendixes

Four appendixes are included at the end of each volume:

- Environmental Timeline, 1620–2004. To understand the history of the environmental movement, each book provides a comprehensive timeline that presents a general overview of activists, important laws and regulations, special events, and other environmental highlights over a period of more than 400 years.
- Endangered List of U.S. Wildlife Species by State.
- Website addresses by classification.
- Government and nongovernment environmental organizations.

Conventional Energy Sources

The United States and many of the industrialized nations consume a large share of the world's total energy resources. The most common or conventional energy resources are fossil fuels. Fossil fuels are used as energy to produce electricity and to operate automobiles, buses, trains, airplanes, and other machines. Approximately 85 to 90 percent of the energy consumed in the United States and the world comes from fossil fuels such as petroleum, natural gas, and coal.

FOSSIL FUELS

Fossil fuels are naturally occurring, *nonrenewable* energy sources, such as coal, petroleum, and natural gas. They are formed in Earth's crust over millions of years by the chemical and physical alteration of plant and animal residues or wastes. Large demands for fossil fuels began in the 18th century during the *Industrial Revolution*, with the invention of gasoline and diesel-powered vehicles, and through the spread of technology and new inventions.

Refer to Volume III for more information on the Industrial Revolution.

Petroleum Resources

Petroleum was the most important energy source in 2000. It accounted for about 40 percent of the world's energy production. In the same year, the United States consumed 26 percent of the world's production of petroleum. As you can see, the U.S. consumes much of the petroleum produced in the world.

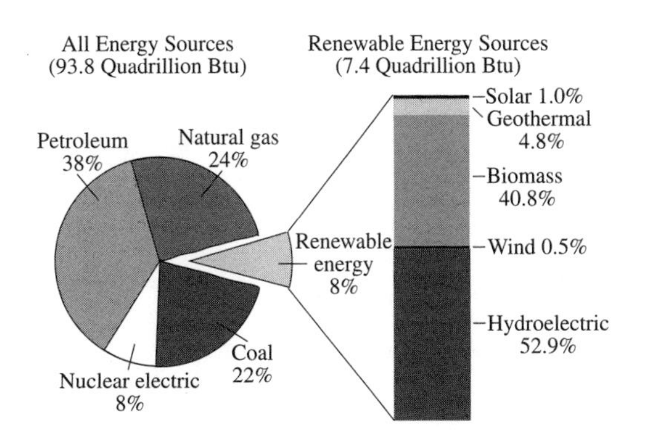

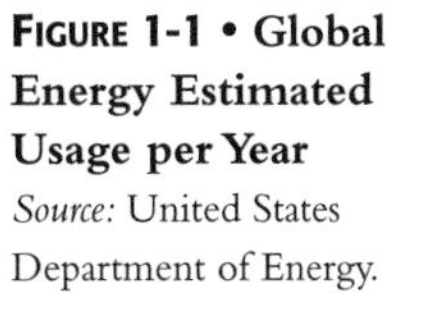
FIGURE 1-1 • Global Energy Estimated Usage per Year
Source: United States Department of Energy.

Energy Basics

Energy is the capacity of a system to do work or a force that produces an activity or causes changes. The system may be a train carrying you and hundreds of passengers across the land, a planet revolving around the sun, or your body growing bone cells. Whether moving or growing, each of these systems is doing work and using energy. Just about everything you see and do involves energy.

Energy exists in many forms. Some of these different forms include

- chemical energy, stored in chemical bonds of molecules
- nuclear energy, found in the nuclear structure of atoms
- electrical energy, associated with the movement of electrons
- thermal energy, associated with the heat of an object (also infrared energy)
- mechanical energy, includes potential energy stored in a system and kinetic energy from the movement of matter (sound is a form of kinetic energy; chemical energy is stored in chemical compounds in fossil fuels such as coal, petroleum, and natural gas, in fuelwood, and in biomass. Water in a reservoir represents potential energy, subject to Earth's gravity)
- electromagnetic energy, associated with light waves such as radio waves, microwaves, x rays, and infrared rays
- radiant energy, associated with human-made sources and the sun (solar energy), both providing thermal energy and light

Humans use a variety of energy sources to do work. These energy resources include chemical (fossil fuels and wood), geothermal, electrical, tidal, hydroelectric, nuclear, solar, and wind energy. One form of energy can be converted to another form. For example, humans convert energy from one form to another when they burn coal. By burning coal, humans release the chemical energy stored in molecules of coal. The chemical energy of the coal is transformed into heat energy.

TABLE 1-1 Greatest Oil Reserves by Country, 2001

2001 Rank	Country	2001 Proved Reserves (Billion Barrels)
1.	Saudi Arabia	261.7
2.	Iraq	112.5
3.	United Arab Emirates	97.8
4.	Kuwait	96.5
5.	Iran	89.7
6.	Venezuela	76.9
7.	Russia	48.6
8.	Libya	29.5
9.	Mexico	28.3
10.	China	24.0

Notes: Figures for Russia are "explored reserves," which are understood to be proved plus some probable. All other figures are proved reserves recoverable with present technology and prices. *Source:* U.S. Energy Information Administration, *International Energy Annual 2000* (May 2002).

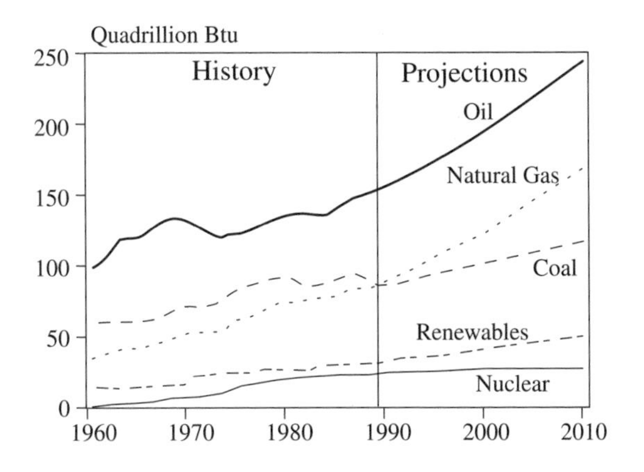

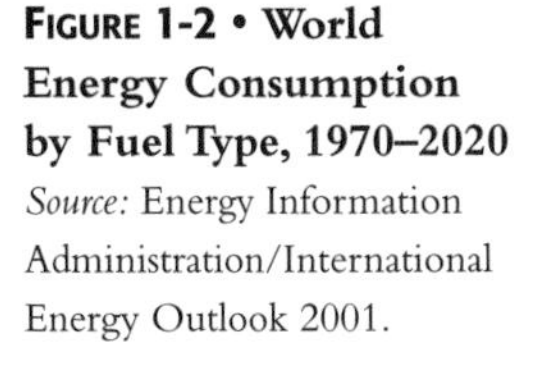
FIGURE 1-2 • World Energy Consumption by Fuel Type, 1970–2020
Source: Energy Information Administration/International Energy Outlook 2001.

Like coal, the use of petroleum dates back thousands of years. Early people used petroleum as a medicine and as a sealant to waterproof their boats. Its use became more common when petroleum was discovered in the United States and served as a fuel for combustion engines in motor vehicles. The big upswing in petroleum growth occurred after World War I. Petroleum, as well as natural gas, began to replace coal as an energy source for home heating and in small industries. By the 1960s, coal was replaced as the world's main source of industrial energy by petroleum and natural gas. Presently, petroleum continues to maintain its position as the world's major source of fuel. According to the Office of Industrial Technologies, every person in the United States consumes an average of 20 pounds of petroleum per day. Gasoline is by far the major petroleum product because of its widespread use in automobiles and other vehicles.

DID YOU KNOW?

In 1859, Edwin Drake constructed the first commercial petroleum well in the United States in Titusville, Pennsylvania. Petroleum was found at 21 meters (68 feet) deep.

PETROLEUM

Petroleum, also known as crude oil, is a combustible, liquid fossil fuel that occurs naturally in deposits, usually underground. The composition of petroleum varies with locality, but it is mainly a mixture of hydrocarbons, of 5 to more than 60 carbon atoms each, with sulfur, nitrogen, and oxygen as impurities. Petroleum is liquid at Earth's surface, varies in density, and is described as heavy, average, or light; light oils are the most valuable because they produce the most gasoline. Petroleum volume is usually measured in barrels, each representing 159 liters (about 42 gallons). One ton of petroleum would equal 7 barrels.

FORMATION OF PETROLEUM Similar in origin to coal, petroleum was formed over millions of years by the chemical and physical alteration of plant and animal remains buried under thick rock layers. Most source rocks of petroleum were deposited in tropical seas; some were later moved and shifted to higher latitudes by continental drift.

Periods of global warming may have accelerated petroleum formation. The late Jurassic, 150 million years ago, was one such period, responsible for the main oil source rocks in the Middle East, the North

Sea, and parts of Siberia. Another occurred in the mid-Cretaceous, 90 million years ago, and was responsible for the oil in northern South America. Much of the oil in the United States comes from older sources, from the Permian period, about 230 million years ago.

MAJOR PRODUCERS OF PETROLEUM The leading producers of petroleum include Russia, Saudi Arabia, Iran, and the United States. However, because the United States consumes about twice as much crude oil as it produces, it must import supplies from other countries. In 2001, the United States produced about 41 percent of its petroleum needs. The rest of the country's petroleum needs was imported from major suppliers that included Canada, Saudi Arabia, Venezuela, Mexico, Iraq, and Nigeria.

Worldwide, the largest reserves are in the Middle East, with the Persian Gulf region producing 27 percent of the world's oil. As of 1993, the United States Geological Survey (USGS) estimated that worldwide, there were 1,103 billion barrels of oil remaining in discovered reserves and 471 billion barrels in undiscovered reserves, almost entirely in oil fields that have already been discovered. Experts believe that the world's supply of petroleum will last 50 to 90 years at the present rate of consumption. Other critics believe the time will be much shorter.

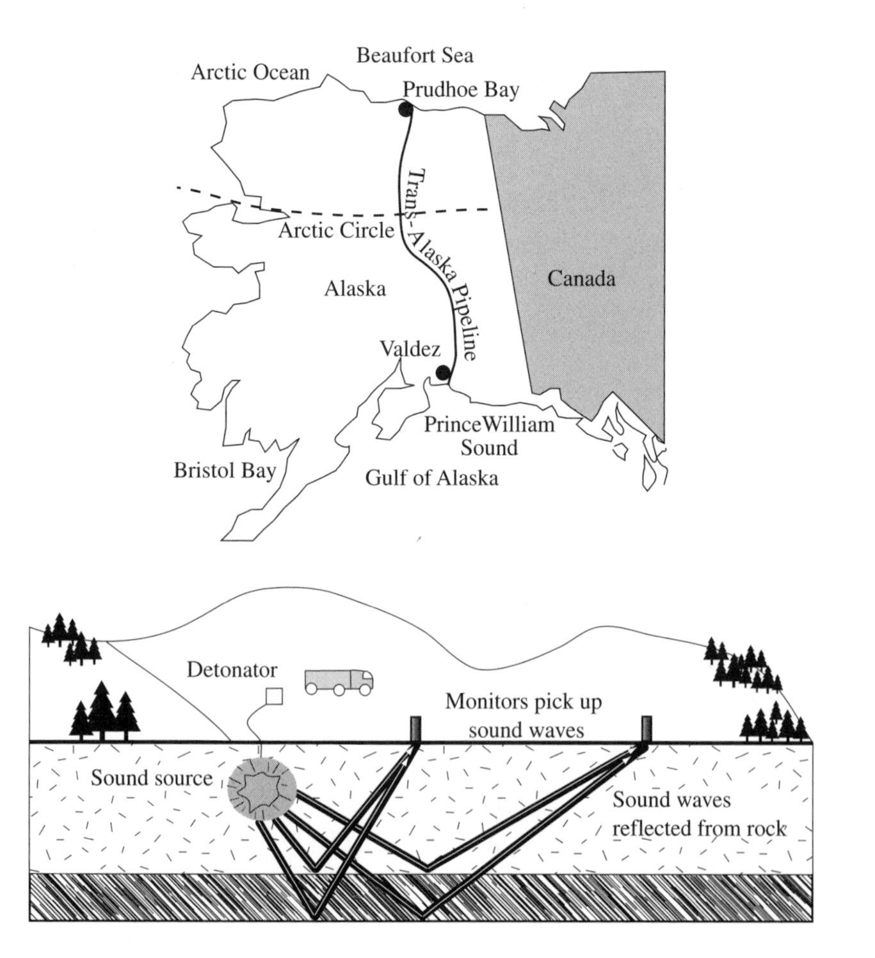

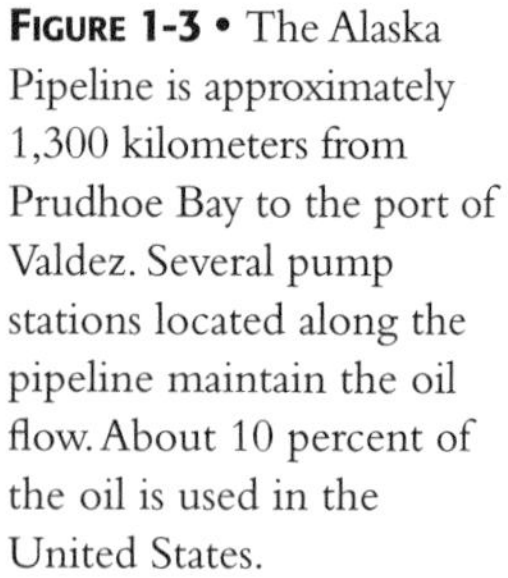
FIGURE 1-3 • The Alaska Pipeline is approximately 1,300 kilometers from Prudhoe Bay to the port of Valdez. Several pump stations located along the pipeline maintain the oil flow. About 10 percent of the oil is used in the United States.

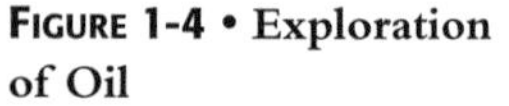
FIGURE 1-4 • **Exploration of Oil**

Petroleum Reservoirs Seismic, magnetic, and gravity surveys are exploration methods used to determine promising sites for oil wells, some of which must be drilled thousands of kilometers deep to reach a deposit. Most petroleum is found in sedimentary rock basins. There are about 700 of these basins worldwide. The sedimentary rocks—sandstone and limestone, generally folded and faulted—are common petroleum reservoir rocks. Oil migrates through porous rock until it is trapped at the top of a fold, against a fault, or where a bed pinches out. Sealing by material such as salt deposits prevents petroleum and natural gas from leaking out of their structural rock traps.

Drilling for Petroleum Although petroleum can seep to the surface, most large reserves are located deep in the ground. To tap these reserves, drilling is necessary. In 1901, the first modern rotary rig was used at the Spindletop oil field, on a salt dome in Texas. The rotary rig used a rotary process. Attached to the rotary drill was a bit with sharp teeth that was rotated by a motor. During the drilling process, the bit cut down into the hard rock as it turned. As the drilling rigs improved, deeper wells were dug reaching deeper deposits. Many deep wells extend more than 4,000 meters (13,000 feet) deep, and others can reach 7,500 meters (24,000 feet) deep or more.

Department of Energy, U.S. (DOE)

A department of the federal government established in 1977 to regulate and manage the energy policy of the United States is the Department of Energy (DOE). The main objective of the DOE is to achieve efficiency in energy use while maintaining environmental quality and a national defense. The DOE is a world leader in the research and development of programs and technologies that generate energy from fossil fuels, nuclear fuels, and alternative energy resources such as solar energy, wind power, and biofuels.

Processing the Petroleum from the Well Once the bit reaches the petroleum, the process of removing the oil from the well begins. Motor-driven pumps are used to lift the petroleum out of the ground. In the pump system, an electric motor drives a gear box that moves a lever. The lever pushes and pulls a polishing rod up and down. The polishing rod is attached to a suckerrod, which is attached to a pump. This system forces the pump up and down, creating a suction that draws oil up through the well. In some cases, the oil may be too heavy

DID YOU KNOW?

The Russian empire led the world in oil production at the turn of the 20th century. By 1919, however Mexico outpaced Russia in oil production because the outbreak of World War I and the revolution in Russia curtailed Russian oil output.

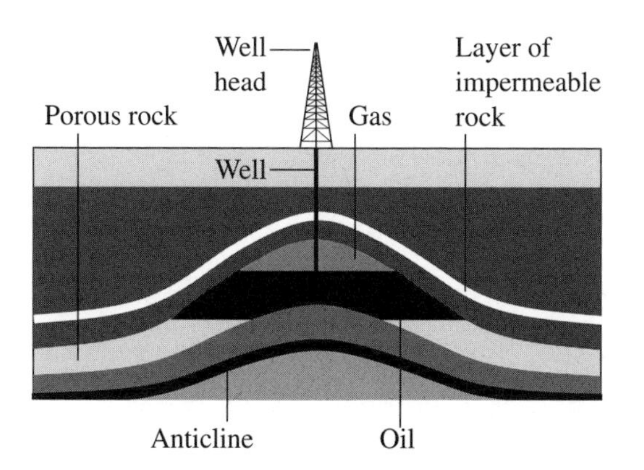

Figure 1-5 • Extracting Oil An oil well is a long shaft that is drilled into rock to obtain oil. The oil is pumped up the shaft from the petroleum deposits below. Several wells are drilled into an oil deposit. Natural gas is generally produced in association with petroleum or with gas wells.

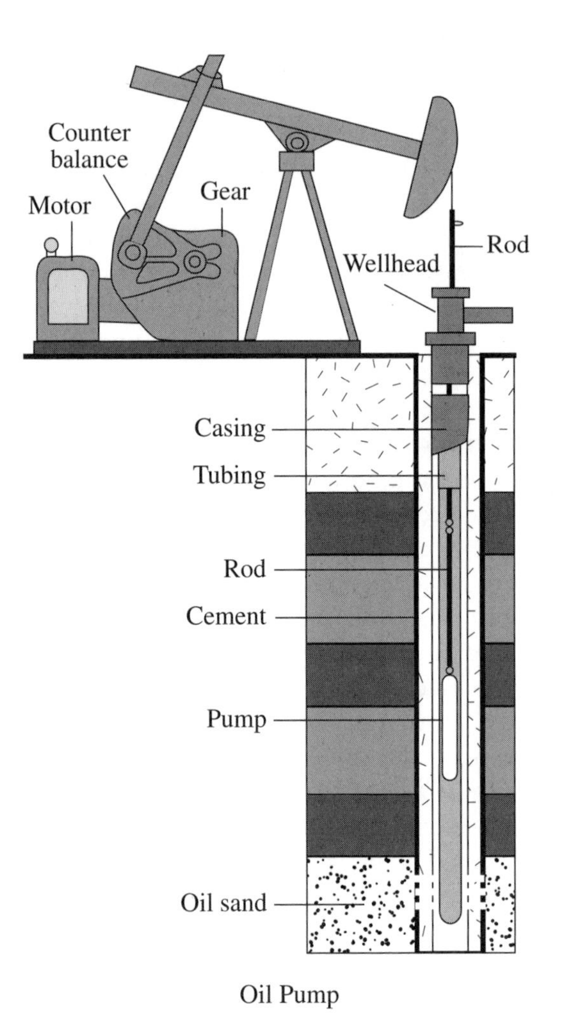

FIGURE 1-6 • Motor-driven pumps are used to lift the petroleum out of the ground.

to flow. A second hole is then drilled into the petroleum reservoir and steam is injected under pressure. The heat from the steam thins the oil in the reservoir, and the pressure helps push the oil up the well. The process by which petroleum or natural gas is removed from a well is called enhanced oil recovery.

Once the petroleum is extracted from the well, it cannot be used in its crude form. The petroleum is transported to a refinery where the petroleum and natural gas are turned into useful products such as gasoline and fuel oil. In a process called *fractional distillation*, the various hydrocarbons in the oil are separated according to their boiling points. The lightest hydrocarbons, such as gasoline, boil off first.

GASOLINE

Gasoline is a light, volatile, highly flammable mixture of hydrocarbons obtained in the fractional distillation of petroleum and used as a fuel for internal-combustion engines and as a solvent. Gasoline is a complex mixture, containing hundreds of different hydrocarbons, most with 5 to 12 carbon atoms per molecule, but varying widely in structure. It is perhaps the most widely used product refined from petroleum. Gasoline is useful as an automobile fuel because it easily evaporates to a gas, which when burned, releases a great deal of energy.

The anti-knock quality of gasoline used in engines is rated by *octane* number. To increase octane rating, additives containing lead were widely used until the late 1960s. Health and environmental hazards were posed by lead, and it created harmful effects. In the 1970s, manufacturers began to change automobile designs and gasoline composition to exclude lead. In 1990, the Clean Air Act (CAA) forced major compositional changes in gasoline; as a result, lead additives are now banned in the U.S. Compounds such as methyl tertiary butyl ether (MTBE), which raises octane ratings and promotes more thorough combustion, and ethanol, used in gasohol, are now added to reduce pollution.

DID YOU KNOW?

Gasoline can also be processed from shale oils and coal. For example, during World War II, Germany produced gasoline from coal. The costs of using this process was very expensive, however.

MTBE (Methyl Tertiary Butyl Ether)

MTBE is an oxygenate additive in gasoline that was developed by petroleum companies in the 1970s to replace lead additives and reduce automobile exhaust emissions. The use of MTBE is one of several methods used to reformulate gasoline for lower emissions. MTBE has improved gasoline combustion and has reduced smog and air pollution in large cities where there are an exceptional number of automobiles, such as Los Angeles County, California. However, although MTBE is making the air cleaner, it is becoming an environmental problem in groundwater that supplies many municipalities with drinking water because MTBE is highly soluble and mobile in groundwater. The danger to humans is unknown, but studies have linked MTBE to liver and kidney tumors in mice. MTBE is being discontinued or phased out as a gasoline additive in the United States.

Other Products

Although gasoline is one of the major products of a refinery, other products can be produced from the fractional distillation. After the gasoline is boiled off at the refinery, the next lightest hydrocarbons are boiled off and collected. Another product includes liquefied petroleum gas (LPG) or bottle gas. This is petroleum gas that is liquefied by pressure and stored in metal tanks. LPG is used in many parts of the world as a main fuel for cooking and heating. Many outdoor home grills use LPG tanks. Other major refinery products include kerosene, diesel fuel, and fuel oil. Kerosene is used as a fuel in jet engine. Diesel fuel is used in locomotives, cars, ships, and trucks. Number 2 heating oil, slightly heavier than diesel fuel, is used to heat homes and buildings and as fuel for electrical-generator power plants that produce electricity. Heavier grades of fuel oil are often used for industrial heat and power because of their greater BTU (energy) value.

Petrochemicals

The chemical industry has made thousands of new chemicals, called petrochemicals, from petroleum. Petrochemical products are used in just about every industry today, from agriculture to medicine. A petrochemical is an organic compound derived from petroleum or natural gas. Petrochemicals are obtained when crude oil or natural gas is refined, or separated, into gasoline, heating oil, asphalt, and other useful substances. Some petrochemicals, such as fuels, dry cleaning fluids,

TABLE 1-2 Products Made from Oil

Ink	Dishwashing liquids	Paint brushes	Telephones
Toys	Unbreakable dishes	Insecticides	Antiseptics
Dolls	Car sound insulation	Fishing lures	Deodorant
Tires	Motorcycle helmets	Linoleum	Sweaters
Tents	Refrigerator linings	Paint rollers	Floor wax
Shoes	Electrician's tape	Plastic wood	Model cars
Glue	Roller-skate wheels	Trash bags	Soap dishes
Skis	Permanent press clothes	Hand lotion	Clothesline
Dyes	Soft contact lenses	Shampoo	Panty hose
Cameras	Food preservatives	Fishing rods	Oil filters
Combs	Transparent tape	Anesthetics	Upholstery
Dice	Disposable diapers	TV cabinets	Cassettes
Mops	Sports car bodies	Salad bowls	House paint
Purses	Electric blankets	Awnings	Ammonia
Dresses	Car battery cases	Safety glass	Hair curlers
Pajamas	Synthetic rubber	VCR tapes	Eyeglasses
Pillows	Vitamin capsules	Movie film	Ice chests
Candles	Rubbing alcohol	Loudspeakers	Ice buckets
Boats	Ice cube trays	Credit cards	Fertilizers
Crayons	Insect repellent	Water pipes	Toilet seats
Caulking	Roofing shingles	Fishing boots	Life jackets
Balloons	Shower curtains	Garden hose	Golf balls
Curtains	Plywood adhesive	Umbrellas	Detergents

solvents, pesticides, drugs, and cosmetic preparations are put to direct use. Most petrochemicals, however, serve as raw materials in the production of *synthetic* substances, particularly plastics.

ETHYLENE A highly reactive gas, ethylene is perhaps the most widely used petrochemical. It's used in the production of plastics, synthetic fibers, and antifreeze. Polyethylene is the most popular plastic material in the world. It is used to make grocery bags, plastic bottles, and children's toys. Polystyrene is a hard plastic that is used for plastic drinking cups, as insulation, and in the manufacture of many kitchen appliances.

VINYL CHLORIDE Another petrochemical is vinyl chloride, an organic gaseous compound consisting of carbon, chlorine, and hydrogen. Vinyl chloride is used to make polyvinyl chloride (PVC), a component of a variety of plastic products, including pipes, wire and cable coatings, packaging materials, furniture, automobile upholstery, wall coverings, housewares, and automotive parts. Until the mid-1970s, vinyl chloride was also used as a coolant, as a propellant in aerosol spray cans, and in some cosmetics. It is no longer used for these purposes because the Department of Health and Human Services (DHHS), the International Agency for Research on Cancer, and the U.S. Environmental Protection Agency (EPA) have all determined that vinyl chloride is carcinogenic to humans.

Other important petrochemicals include benzene, which is used to make synthetic rubber and latex paints, and phenols, important chemicals used in the manufacture of perfumes, artificial flavorings, and pesticides.

Environmental Concerns of Petroleum

The burning of petroleum products and other fossil fuels is responsible for approximately 80 percent of the world's carbon dioxide emissions, 25 percent of U.S. methane emissions, and 20 percent of global nitrous oxide emissions. Processing of petroleum and use of its products also create many other air pollutants, including airborne particulate matter. About 20 percent of all U.S. greenhouse gas emissions are attributable to motor vehicle gasoline consumption. And about 60 percent of the total weight of pollutants discharged into the atmosphere originates from this source.

Unfortunately, the production and use of petrochemicals causes a variety of environmental problems. When these substances are produced, for example, a number of pollutants, including sulfur dioxide and particulates, are released into the air. Emission of sulfur dioxide is one of the main contributors of *acid rain* formation. Certain petrochemicals themselves, such as benzene and toluene, are also highly toxic to humans and other organisms.

In the United States, the EPA tightly monitors the safety of petrochemical plants. Strict environmental laws, including the Clean Air Act, help control the amount of pollutants released by petrochemical plants. However, there will always be a tradeoff between the environmental dangers produced by petrochemicals and the need of people and society to use these products.

For more information about the impact of petroleum and petrochemical products on the environment, refer to Volume IV.

Natural Gas

In 2000, dry natural gas was the third primary energy source, just shy of tying coal for the second spot on the world's fossil fuel production list. Natural gas (NG) consists mainly of methane, the simplest hydrocarbon.

Natural gas has been around for some time. The Chinese used natural gas as an energy source more than 3,000 years ago. They ignited the natural gas to produce heat to evaporate pools of brine water to make salt. To extract the natural gas, the Chinese dug wells that extended about 500 meters (1,500 feet) deep. To transport the gas to the surface, the Chinese used pipes made of bamboo.

TABLE 1-3 Greatest Natural Gas Reserves by Country, 2001

2001 Rank	Country	2001 Proved Reserves (Trillion cu ft)
1.	Russia	1,700.0
2.	Iran	812.3
3.	Qatar	393.8
4.	Saudi Arabia	213.3
5.	United Arab Emirates	212.1
6.	United States	177.4
7.	Algeria	159.7
8.	Venezuela	146.8
9.	Nigeria	124.0
10.	Iraq	109.8

Notes: Figures for Russia are "explored reserves," which are understood to be proved plus some probable. All other figures are proved reserves recoverable with present technology and prices. *Source:* U.S. Energy Information Administration, *International Energy Annual 2000* (May 2002).

Liquid Natural Gas

In 1999, many taxi and bus drivers in Cairo, Egypt, converted their engines to run on liquid natural gas (LNG). This technology uses LNG, which is compressed natural gas. Cairo is the world leader in the number of privately owned gas-powered motor vehicles, and now the country's bus and taxi companies are coming aboard. Egypt has abundant natural gas reserves and can offer LNG car owners a fuel that is a less expensive and cleaner burning fuel than gasoline. LNG vehicles produce about 80 percent less carbon monoxide and fewer hydrocarbons than gasoline-powered vehicles. And LNG costs less than gasoline in Egypt. A cubic meter of LNG is 50 percent less expensive than the equivalent amount of gasoline sold in Cairo. By 2010, Egypt plans to have more than 25 stations to service LNG vehicles.

Today residential and commercial uses consume the largest proportion of natural gas in North America and Western Europe. In these areas, natural gas is commonly used for home heating and cooking. Gaseous fuels are convenient to use because they can be readily turned on and off, produce no smoke, and leave no ash behind. After residential users, industry is the next largest consumer, and electric-power generation is third. Natural gas is also used to fuel some automotive vehicles, though currently on a very limited basis in some countries. Experts believe that gas reserves will last 60 to 100 years at the current rate of consumption.

LOCATING NATURAL GAS DEPOSITS

Natural gas is often found in solution with petroleum, in sand, sandstone, and limestone deposits. Natural gas was formed in Earth's crust over millions of years by the chemical and physical alteration of organic matter. In order for gas to accumulate, it must be trapped. The underground gas reservoir must be sealed at the top by an impermeable stratum or cap rock, such as clay or salt. The entire cover structure must be shaped in such a way as to prevent gas from leaking to the surface. Gas accumulations are mostly encountered in the deeper parts of sedimentary basins. On the Gulf Coast of the United States, for example, more than half of the deposits discovered at depths greater than

Propane is gas found in petroleum, which is sold in liquid form under pressure as LPG. Propane buses are used in Zion National Park. (Courtesy of Hollis Burkhart.)

3,600 meters (10,000 feet) are gas fields. Among the largest accumulations of natural gas are those of Urengoy in Siberia, the Texas Panhandle in the United States, the Slochteren-Groningen area in the Netherlands, and Hassi R'Mel in Algeria.

In the United States, natural gas was first used to light the town of Fredonia, New York, in 1821; however, the fuel's use remained localized over the next century because long-distance transportation of gases was difficult. Methods of pipeline transportation were developed in the 1920s; the period between World War II and the 1980s saw tremendous residential and commercial expansion that relied increasingly on the use of pipeline transportation of gas. North American gas pipelines now extend from Texas and Louisiana to the Northeast Coast, and from the Alberta gas fields to the Atlantic seaboard.

The Contents of Natural Gas

Natural gas is a mixture of flammable gases including methane, ethane, propane, and butane. Since natural gas has no smell, a substance is added to the natural gas to produce an odor that enables us to detect a gas leak. The mixture is usually composed of 70–80 percent methane. Other hydrocarbon constituents include ethane and propane, which are used as nonrenewable fuels. However, the composition of natural gas varies according to locality; minor components may include carbon dioxide, nitrogen, hydrogen, carbon monoxide, and helium. Natural gas is the cleanest-burning of the fossil fuels, yielding little more than carbon monoxide, carbon dioxide, and water as combustion products. Although some natural gases can be used directly from the well without treatment, most must first be processed to remove undesirable constituents such as hydrogen sulfide and other sulfur compounds.

Coal Resources

Coal is the second primary energy source, accounting for 23 percent of the world's energy needs. China, the United States, India, Australia, and South Africa produce much of the world's need for coal.

DID YOU KNOW?

Coal mining in Pennsylvania fueled the Industrial Revolution in the United States in the mid-1700s.

Coal has been used for thousands of years. Archeological evidence shows that China was burning coal in 1100 B.C. But not until brick chimneys became popular did people burn coal indoors. During the Industrial Revolution in England, the common use of steam engines led to a surge in the demand for coal. Approximately 100,000 coal-fed steam engines were used to power machinery, trains, and steamboats and for pumping water out of coal mines.

Today coal, the most abundant fossil fuel in the world, is used primarily to produce electricity and to a lesser degree produce heat for buildings. Coal is a burnable carbonaceous rock, a rock that contains organic matter (plant and animal residues) and carbon. Besides carbon, there are other elements in coal, including hydrogen, oxygen, sulfur, and nitrogen. But coal also has various amounts of mineral matter. Therefore, coal may also be considered a mineral of organic origin.

FORMATION OF COAL

DID YOU KNOW?

Coal has been mined for more than 1,000 years, and large-scale mining was practiced as early as the eighteenth century. The first coal mine in America was opened in Virginia in the Appalachian bituminous field during the 1750s; the mining of anthracite began in the late 1700s.

During the carboniferous period of geologic time (280 to 345 million years ago), scientists believe, great quantities of vegetation and other organic matter collected and underwent slow decomposition. This occurred mainly in large shallow swamps, lakes, and marshes and in lagoons where a spongy, brown material called peat was formed. The formation of peat was the first step in the coal-making process. Over time, the peat was compacted beneath other deposits. As a result, water was squeezed out of the peat and gases such as methane were expelled into the atmosphere. Taking millions of years, the continued processes of burial and compression eventually converted the peat to coal. The greater the heat and pressure, the harder the coal. As a result, there are different grades of coal: lignite, bituminous, and anthracite. Lignite is the lowest grade with the highest percentage of *volatile* matter. Bituminous is a step up from lignite and is the most abundant of the three types of coal. Anthracite, or hard coal, is the highest-grade coal, with a high carbon content and a low percentage of volatile matter.

FOUR MAJOR TYPES OF COAL

The four major types of coal are measured on heating value expressed in British Thermal Units, or BTUs, per unit weight. One BTU is the heat required to raise one pound of water one degree Fahrenheit, about 253 calories.

Lignite. A brownish-black coal with generally high moisture and ash content and the lowest carbon content and heating value. BTU/lb: 5,500 to 8,000.

Sub-Bituminous. A dull black coal with a higher heating value. BTU/lb: 8,000 to 12,000.

Bituminous. A soft, intermediate grade of coal that is the most common and widely used coal in the United States. BTU/lb: 11,000 to 15,000.

Anthracite. The hardest type of coal, consisting of nearly pure carbon. Anthracite has the highest heating value and lowest moisture and ash content. BTU/lb: 13,000 to 16,000.

Uses of Coal

Coal is used primarily for the generation of electricity. In fact, coal generates 56 percent of the electricity consumed in America each day, far more than any other energy source. According to the World Coal Institute, about 36 percent of the world's electricity is produced by burning coal. Coal is a major fuel for generating electricity in Poland (97 percent of electricity), South Africa (93 percent), Australia (85 percent), China (80 percent), and India (75 percent). During the next decade, coal use is expected to rise in Southeast Asia, where coal will be the major fuel for producing electricity. Besides generating electricity, coal is also used to produce steam for heating buildings, in the manufacturing process of iron and steel, and for *metallurgical* operations.

DID YOU KNOW?

Today, in West Virginia, 99 percent of the state's electricity comes from coal.

Locations of Coal

Coal deposits are found all over the world. Even the Antarctica has coal deposits. But most of the coal reserves are found in large deposits in the mid-latitudes of the Northern Hemisphere. There are few coal deposits in the Southern Hemisphere. In all, about 100 countries have coal reserves.

Most of the world's largest deposits are in North America, Eastern Europe, Russia, China, India, and Africa. In the United States, coal is

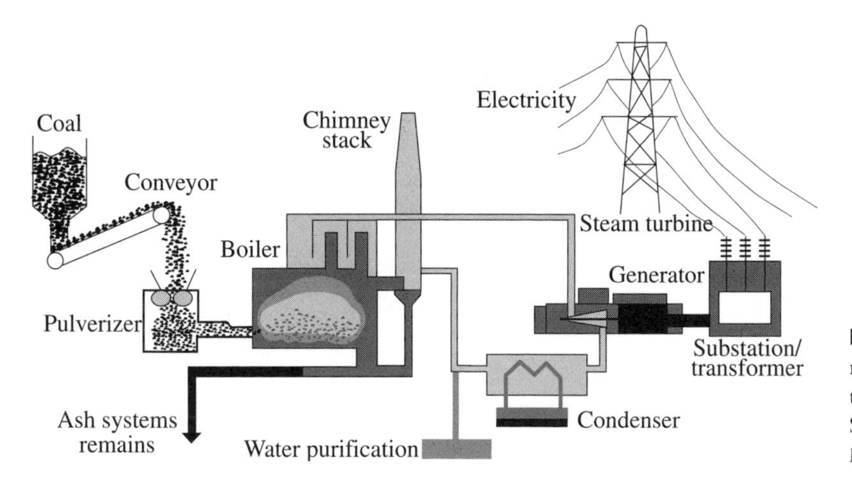

Figure 1-7 • Coal generates more than half the electricity used in the United States. *Source:* National Mining Association.

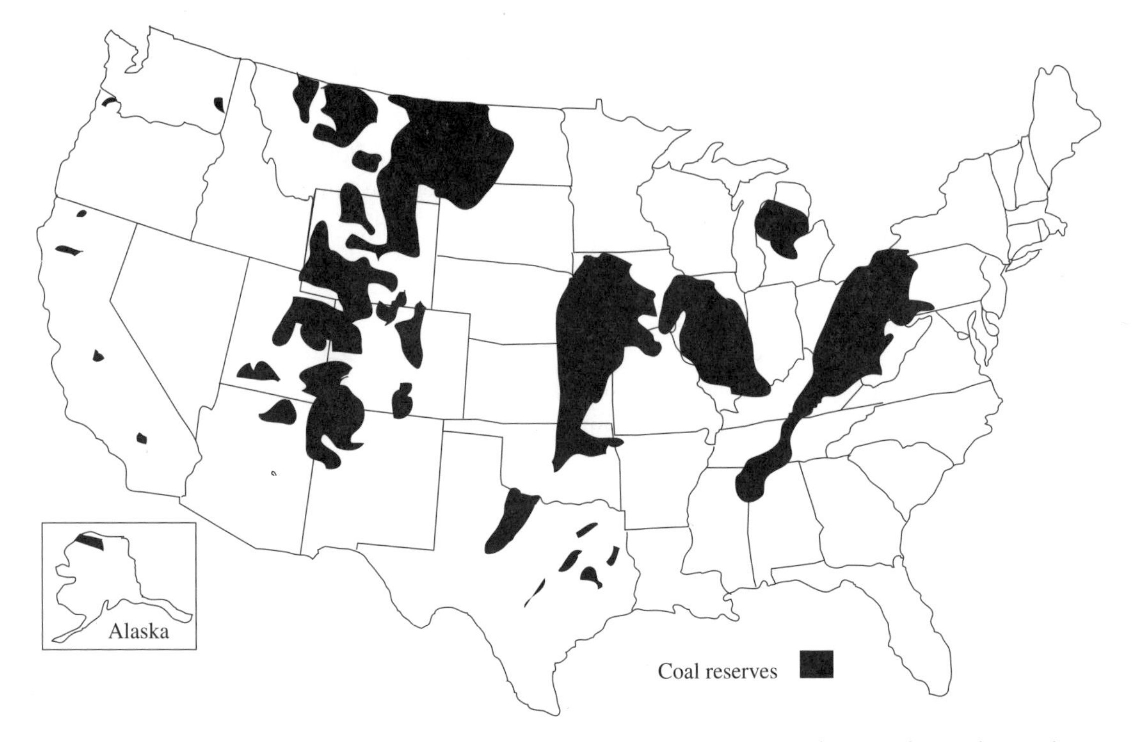

FIGURE 1-8 • Coal Reserves in the United States The United States has nearly 275 billion tons of recoverable coal. That's more than 250 years of supply at today's usage rates. The largest coal producing state is Wyoming, with 373 million tons of production 2002. *Source:* National Mining Association.

found in 38 states and nearly one-eighth of the country lies over coal beds. The top coal-mining states include Montana, Illinois, Wyoming, West Virginia, Kentucky, Pennsylvania, Ohio, Colorado, Texas, and Indiana. Recent estimates indicate that the world's supply of coal should last for another 250 to 400 years at current production and consumption levels.

MINING OF COAL

According to the American Coal Foundation, there are two basic ways to mine coal—underground mining and surface mining.

UNDERGROUND MINING Underground mining is used to extract coal lying deep beneath Earth's surface or in seams exposed on hillsides. The coal is reached by drilling two openings into the coal bed to transport workers and equipment and to send coal to the surface. Both openings serve to circulate air in the mine. Coal is broken up and mined by one of several methods:

Conventional mining. This is the older practice of using explosives to break up a coal seam.

Continuous mining. A huge machine with a large rotating steel drum equipped with tungsten carbide teeth scrapes coal from a seam at high speeds.

Long wall mining. A cutting machine with a large rotating steel drum is dragged back and forth across a "long wall" or seam of coal. The loosened coal falls onto a conveyer belt for removal from the work area. After coal is extracted, it is removed by automatic extraction systems that cut the coal. The cut coal is then loaded onto shuttle cars to a central loading area in the mine or placed on belt conveyors that remove coal to the surface. Railroad cars are used to transport the coal long distances and to docks where the coal is loaded onto barges for shipment overseas.

Surface Mining Surface mining, also called strip mining, is used when coal is found close to the surface or on hillsides. Compared to subsurface mining, surface mining generally costs less, is safer for miners, and usually results in the removal of a greater percentage of coal or other minerals underground.

Surface Mining of Coal. Surface mining is the removal of coal and minerals from regions located at or near Earth's surface. Approximately two-thirds of todays coal production results from surface rather than underground mining in the United States. (Courtesy of Tom Repine, West Virginia, Geological and Economic Survey.)

Surface mining essentially involves removing the *overburden*, which covers the coal seam, using heavy earth-moving equipment such as strip shovels and bulldozers. Surface mining may involve digging up 10 to 25 meters (33 to 82 feet) of topsoil and rock to reach the coal. The topsoil and the overburden are carried away and stored, to be used again once all the coal is removed.

Once the coal is scooped out of the coal seam, it is broken up and shoveled directly out of the ground. As more digging occurs, an open pit develops. Once all the coal is removed, the overburden is placed back in the pit. The topsoil is spread over the overburden and seeded so plants and vegetation will again grow in the former mining area. Replacing the excavated soil and reestablishing vegetation and plant life is a process known as reclamation. Reclaimed land has been successfully used for wildlife preserves, golf courses, recreational parks, sites for commercial development, pasture land, native habitats, and productive farmland.

DID YOU KNOW?

Coal from the United States is exported to more than 40 countries around the world. Currently, Canada, Japan, and Italy are among the biggest customers receiving shipments of U.S. coal.

Much of the reclamation activity in the United States has been the result of legislation passed in 1977, called the Surface Mining Control and Reclamation Act. The act requires companies involved in surface mining operations to restore mined lands back to their natural conditions after the mining operations cease. It also prohibits surface mining on certain lands, such as national forests. However, not all countries have reclamation programs.

Environmental Concerns of Coal Environmental issues associated with coal uses for energy include air pollution from coal-fired power plants and the impact of coal mining on landscapes. Coal is the dirtiest of all fossil fuels and produces major air pollutants when it burns. Coal-fired plants emit pollutants such as carbon dioxide (CO_2), sulfur dioxide (SO_2), and nitrogen oxides (NO_x) into the atmosphere. Studies indicate that about 70 percent of all sulfur dioxide and 35 percent of carbon dioxide emissions pumped into the atmosphere come from coal-burning power plants. Other air pollutants include *volatile organic compounds (VOCs)*, soot, ash, and other particulate matter. Heavy metals such as cadmium (Cd) and mercury are also released from coal-burning plants. Such plants also produce bottom ash and fly ash that need to be collected, recycled, or disposed of in landfills.

Refer to Volume IV to learn more about the impact of the use of coal on the environment.

Underground mines and open-pit mines can be a serious environmental problem if left abandoned. Acid mine drainage (AMD) is a water pollution problem resulting from the discharge of acidic water from coal or other mines containing mineral ores for iron, copper, lead, or zinc into streams and rivers. AMD also results when rainwater leaches through overburden or tailings—the waste materials produced by mining operations.

Electricity

Electricity is a form of energy that accounts for almost 40 percent of all the energy use in the United States. Globally, electricity is generated mostly by fossil fuels. Besides fossil fuels, geothermal, nuclear, water power, wind, and solar energy are used to produce electricity, but on a smaller scale.

Electricity is electron energy that can be used to produce light, mechanical energy, or heat. The flow of electrons through a conductor such as metal or wire is referred to as an electric current. The rate at which the electric charge flows in the current is measured in amperes. Current is measured with an ammeter. The complete route of an electric current starts from a source such as the terminal of a battery or generator through the resistance of an electric-powered mechanism such as a toaster and returns back to the other terminal completing the electric circuit. The electric circuit is a closed path through which electrons can flow.

The rate at which electric energy is converted to another form of energy is called electric power. The most common unit of electric power is the kilowatt (kW), which is the unit of power equal to 1,000 watts (1.34 horsepower) or the energy consumption rate of 1,000 joules per second. Electric utility companies charge customers a rate based on the number of watts the customers use. The standard measure for large amounts of electricity is the kilowatt-hour (kWh), which equals 1,000 watts per hour. A lit 100-watt bulb would consume one kilowatt in an hour. The average home in the United States uses 7,500 to 9,000 kilowatt-hours in a year's time. Electric utility companies use billing rates based on cents per kilowatt-hours.

Here are the average power requirements of some home appliances: hair dryer (600 to 1,000 watts), microwave (1,300–1,450 watts), color television (200–300 watts), oven range (2,600 watts), and a refrigerator (500–700 watts). Energy costs in a home can be calculated by multiplying the average power in watts (w) of an appliance by the total hours of use for a month. The total will give monthly watt hours, which can be converted to kilowatts per hours (kWh) per month. Then to get monthly cost, multiply kWh times the utility rate (dollars per kWH) charged by the electric utility company.

TABLE 1-4 Power Requirements of Home Appliances

Appliance	Power (W)	Typical Energy Consumption (kWh/year)
Clock	2	15–50
Electric clothes dryer	4,600	900–1,000
Hair dryer	1,000	60
Light bulb,	100	108
compact fluorescent	18	19
Refrigerator	360	1,600
Television	350	300–1,000
Washing machine	700	1,008

Vocabulary

Acid rain Acid precipitation in the form of rain or snow that contains a higher level of acid than normal rain. Acid rain is usually caused by sulfur dioxide and nitrogen dioxide from the burning of fossil fuels.

Fractional distillation A distillation process where the parts of a mixture are collected at different phases of the distillation.

Industrial Revolution A period of social and economic change beginning in the middle of the eighteenth century in England; began later in the United States.

Metallurgical Refers to the extraction of metals from their ores.

Nonrenewable Natural resources such as fossil fuels which cannot be replaced if they are consumed.

Octane Hydrocarbons found in gasoline. Octane rating refers to the quality and performance of gasoline according to the combustion of hydrocarbons in it.

Overburden Earth and rock that covers the coal layer or seam.

Synthetic Human-made materials usually produced from petrochemicals.

Volatile A liquid that evaporates into a gas rapidly at room temperatures.

Volatile organic compounds (VOCs) Carbon hydrogen compounds that represent a major category of air pollutants.

Activities for Students

1. Using Table 1-2 on page 8, determine how many of them are used in your home. Which does your family use the most? Estimate the average daily cost of each kind of energy source.
2. Comparing and contrasting at least two types of fuels described in this chapter, explain why the term "fossil" is used to describe these fuels.
3. Create a chart to show when in history each type of fossil fuel was first used, when it became a main source of energy, and in which countries the resources have been mostly found.
4. Go to the Oil & Gas Journal Online Website, read a selection of the articles, and summarize what you determine to be the journal's purpose and audience.

Books and Other Reading Materials

Burleson, Clyde W. *Deep Challenge!: The True Epic Story of Our Quest for Energy beneath the Sea.* Gulf Professional Publishing Company, 1998.

Kittinger, Jo S. *A Look at Rocks: From Coal to Kimberlite.* New York: Franklin Watts Incorporated, 1997.

Snedden, Robert. *Energy from Fossil Fuels (Essential Energy).* Boston: Heinemann Library, 2001.

Websites

American Gas Association, http://www.aga.org

American Petroleum Institute, http://www.api.org

Oil & Gas Journal Online, http://www.ogjonline.com

U.S. Department of Energy, Energy Information Administration, http://www.eia.doe.gov

U.S. Department of Energy, Office of Fossil Energy, http://www.fe.doe.gov

U.S. Geological Survey Energy Resources Program, http://energy.usgs.gov/index.html

Nuclear Energy

Today most of the industrialized countries rely heavily on fossil fuels for the production of their electricity needs. Coal, natural gas, and petroleum are used in power plants to heat up and convert water to steam, which drives a turbine to generate electricity. However, there are about 30 or so countries that also use nuclear-powered energy plants to produce electricity.

WORLD USE OF NUCLEAR ENERGY

According to the World Nuclear Association, as of 2001, there were 438 nuclear power reactors worldwide. On the global scene, some 35 countries, including the United States, have chosen nuclear power as part of their energy needs. About 16 percent of all the world's electricity is produced from nuclear power plants. France is a major global producer of nuclear power for electricity. About 80 percent of France's electricity is produced by nuclear energy. The country is a major exporter of electricity to other countries.

As of 2001, the United States had 104 nuclear power plants that provide about 20 percent of its electricity needs. Six states rely on nuclear power for more than 50 percent of their electricity, and another thirteen states rely on nuclear power for up to 25 to 50 percent of their electricity.

NUCLEAR ENERGY BASICS

Nuclear energy is referred to as an alternative energy source that might replace fossil fuels in the future. Nuclear energy is stored within the nuclei of *atoms* of uranium, a nuclear fuel. The energy may be released through a nuclear process called fission. Fission means to split apart. In nuclear fission, a heavy nucleus of an atom absorbs a neutron and splits into two lighter *nuclei* forming new elements and releasing several neutrons plus energy. These nuclear reactions produce much

DID YOU KNOW?

Another process for producing nuclear energy is fusion, which is the combining of a nuclei, such as hydrogen, into a heavier nuclei. Fusion occurs within the core of the sun where the pressure and temperature are extremely high. However, using nuclear fusion presently as an energy source is still in the experimental stages.

For more information on fusion, see end-of-chapter text.

TABLE 2-1 World Use of Nuclear Energy

Rank	Country	Electricity from Nuclear Generators (% of Total)
1	Lithuania	77
2	France	76
3	Belgium	55
4	Sweden	46
5	Ukraine	45
6	Slovakia	44
7	Slovenia	38
8	Bulgaria	42
9	Republic of Korea	41
10	Switzerland	41

more energy per unit of fuel weight than is produced with conventional materials such as coal or petroleum. For example, when one kilogram of uranium undergoes fission, it releases energy equal to that released by the burning of 6,000 metric tons of coal or 18,000 metric tons of TNT explosive.

NUCLEAR POWER PLANTS

In the 1950s, utility companies began to build nuclear power plants to produce electricity. The world's first commercial-scale nuclear reactor power plant began to operate in Britain in 1956. In the United States, the first licensed atomic or nuclear power plant opened in November 1957 in Vallecitos, California.

How Nuclear Reactors Function

The main component of a nuclear power plant is the nuclear reactor in which a *chain reaction* can be started and controlled over a period of time. Chain reactions in a nuclear reactor generally begin with a nuclear fission of an enriched uranium oxide (U-235) nucleus, the most common nuclear fuel. In a chain reaction, a neutron bombards the nucleus, causing it to break apart and release nuclei of other *elements*, free neutrons, and energy. The free neutrons then strike other U-235 nuclei, causing them to also undergo fission. The free neutrons continually cause more

Nuclear energy can be converted to other form of energy. Atoms in the nuclear fuel are split, releasing their nuclear (mass) energy and creating thermal energy. This heat energy is, in turn, captured in the form of steam and used to drive a turbine generator, creating kinetic energy. The kinetic energy spins a magnetic field around a conductor, causing a current to flow—creating electrical energy.

History of Using Nuclear Energy

The first fission reaction was achieved in 1939. In 1942, it was experimentally proven that a self-sustaining chain reaction could be produced in Uranium 235 or U235 and plutonium-239. Because these discoveries were made during World War II, nuclear energy was used for the destructive purpose of creating the atomic bomb, in which fission progresses rapidly to produce an explosion. Continuous production for many months was necessary to produce the quantities of materials required to build the bomb; the Manhattan Project was a top-secret program of the U.S. government conducted between 1942 and 1946 for the purpose of developing an atomic (nuclear) bomb for use during World War II. To accomplish this goal, several secret laboratories, overseen by the U.S. Army, were established. Primary among these were the Metallurgical Lab in Chicago, Illinois (now known as the Argonne National Laboratory–East), for nuclear reactor research; the Clinton Engineer Works (now known as the Oak Ridge National Laboratory), for isotope separation; the Hanford Reservation of Washington (now the Hanford Nuclear Waste Site), for plutonium-239 production; and the Los Alamos National Laboratory, for bomb assembly.

As a result of the Manhattan Project, three atomic bombs (*Trinity Test, Fat Man,* and *Little Boy*) were produced at a cost of more than $2 billion. In August 1945, two atomic bombs were dropped on the cities of Hiroshima and Nagasaki, Japan. The explosive power of each bomb was approximately equal to that of 18,000 metric tons of TNT. The production and use of the bombs brought an end to World War II and marked the beginning of the construction and use of nuclear weapons, as well as a beginning to finding other uses for nuclear energy.

After World War II, a major effort was made to apply nuclear energy to peacetime uses. Nevertheless, five additional nations—the Soviet Union, the United Kingdom, France, China, and India—soon demonstrated the capability to explode nuclear devices. In 1970, the Treaty on the Nonproliferation of Nuclear Weapons went into effect. Signatory nations without nuclear weapons agreed not to develop them in exchange for the provision of non-nuclear materials and technology from the nations that already had nuclear weapons. In a major effort to limit the nuclear arms race between the United States and the Soviet Union, negotiations such as the Strategic Arms Limitation Talks (SALT) were pursued during the 1980s. The International Atomic Energy Agency (IAEA) attempts to ensure that weapons *proliferation* does not occur.

fission, which releases more neutrons. This leads to a series of fissions with more and more of the nuclei being split apart. Each time the U-235 breaks up, large amounts of heat energy are released. The heat energy in the nuclear reactor is used to boil water. The boiling water evaporates into steam, which drives a turbine. The mechanical energy of the turbine produces electricity.

Components of a Nuclear Reactor

The design of a nuclear reactor varies, but the general features include a thick, reinforced concrete and steel structure called a containment. The containment walls, one-meter (three feet) thick, protect the reactor core, where the fission takes place. Other components of a reactor include fuel, fuel rods, pumps, moderators, coolants, a steam generator, and control rods.

FIGURE 2-1 • Nuclear Power Plants The approximate number of commercial nuclear power plants in the United States as of 1999. Since then some nuclear power plants have been taken offline or closed down.

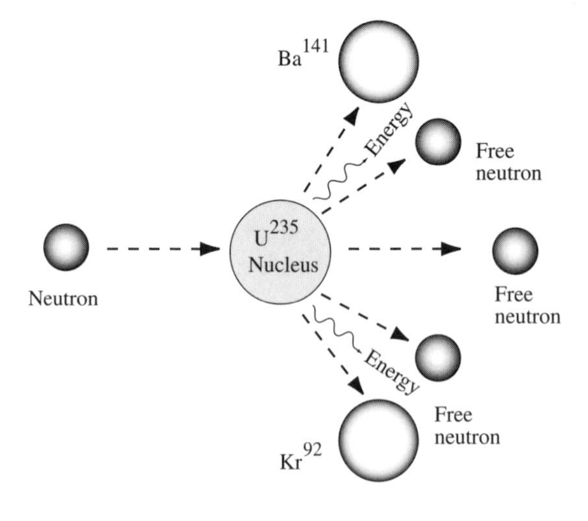

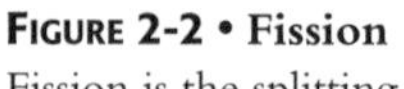

FIGURE 2-2 • Fission Fission is the splitting of an atom into smaller fragments with the release of energy. The process of fission is used in a nuclear reactor to produce energy and in detonating nuclear weapons.

Nuclear power plants have cooling towers. Cooling towers are used to dissipate waste heat into the atmosphere from a coal-fired or nuclear electric power plant and other installations. In the foreground are several rows of photovoltaic panels that are used to provide electricity as well. (Courtesy of Warren Gretz, National Renewable Energy Laboratory.)

NUCLEAR FUEL

Uranium (U), is a radioactive metallic element that is an energy source as a fission fuel in nuclear reactors. Uranium and its byproducts are naturally present in air, soil, and water, as well as in trace amounts in the food we eaten. The most common form of uranium is called U-238, which has a half-life of 4.5 billion years. Uranium (atomic number 92) has natural *isotopes* with atomic masses ranging from 234 to 239 atomic mass units ("amu").

Uranium exists naturally combined with oxygen in minerals such as pitchblende and carnotite. Uranium minerals are widely distributed in Earth's crust. They are present in sandstones, in veins within rock fractures, and in *placer deposits*—ore materials that have been transported and deposited in river deltas and streams. Most uranium mined in the United States derives from sandstone deposits. Worldwide, the richest deposits of uranium are present in France, the Russian Federation, the Ukraine, Australia, Canada, and southern Africa. Some experts believe that there is about a 40-year supply of uranium reserves at the current rate of consumption.

DO YOU KNOW?

Uranium is very dense heavy metal. Density increases weight. As an example, a gallon of water weighs about 8 pounds. The same gallon containing uranium would weigh about 150 pounds!

FUEL RODS

The fuel used in nuclear reactors is natural uranium oxide or U-235, enriched uranium oxide. U-235 is one of the fissionable isotopes of uranium. The enriched fuel is made into pellets and placed inside fuel rods made of a zirconium alloy or other material. The fuel rods are joined together in a *reactor core*. When the U-235 is bombarded with neutrons, fission reaction takes place in the reactor core:

$$_{92}U^{235} + {_0n^1} \rightarrow {_{36}Kr^{90}} + {_{56}Ba^{142}} + \text{neutrons}$$

MODERATOR AND COOLANTS

The moderator in the nuclear reactor is used to slow down the neutrons so that the right speed is maintained for a steady fission rate. The moderators contain a variety of materials that include pure water, heavy water or *deuterium* oxide, and graphite. Coolants are piped into and out of the reactor core removing excessive amounts of heat that build up in the reactor. The steam generator is part of the cooling system in which the heat from the reactor is used to make steam for the turbine.

The list of coolants include pure water or heavy water, graphite, carbon dioxide, sodium, and helium. Sometimes the coolant is also the moderator. The discharges of the heated water are pumped into cooling towers or in nearby waterways.

CONTROL RODS

The chain reaction is regulated by control rods made from neutron absorbing materials such as *cadmium* or boron. The control rods in the reactor core are raised or lowered to speed up, slow down, or stop the fission.

Types of Reactors

LIGHT WATER REACTOR (LWR)

Even though there are several kinds of reactors, most nuclear electricity is generated from just two types of reactors that were developed in the 1950s. They have been modified and improved since that time. The most common type that is used throughout the world is the light water reactor (LWR) or pressurized water reactor (PWR). The LWR design was based on the nuclear submarine power plant. The reactor uses water at a very high pressure in one section of the reactor and in a second section that generates steam to drive a turbine. Today there are about 230 of the light water reactors in the world used for generating electricity.

LWR fuel assemblies contain approximately 200 to 300 fuel rods arranged vertically. The fuel rods contain low-enriched uranium U-235 (3 percent ^{235}U and 97 percent ^{238}U). Each rod holds a fuel pellet that is about 1.0 centimeter (less than 1/2 inch) in diameter and 1.5 cm long. Each rod can extend to about 3.5 meters (about 11 feet) long. These reactors used pure water (H_2O), which act as a moderator and a coolant. Water is boiled in the heat exchanger providing high-pressure steam that drives a turbine to produce electricity. The spent steam is

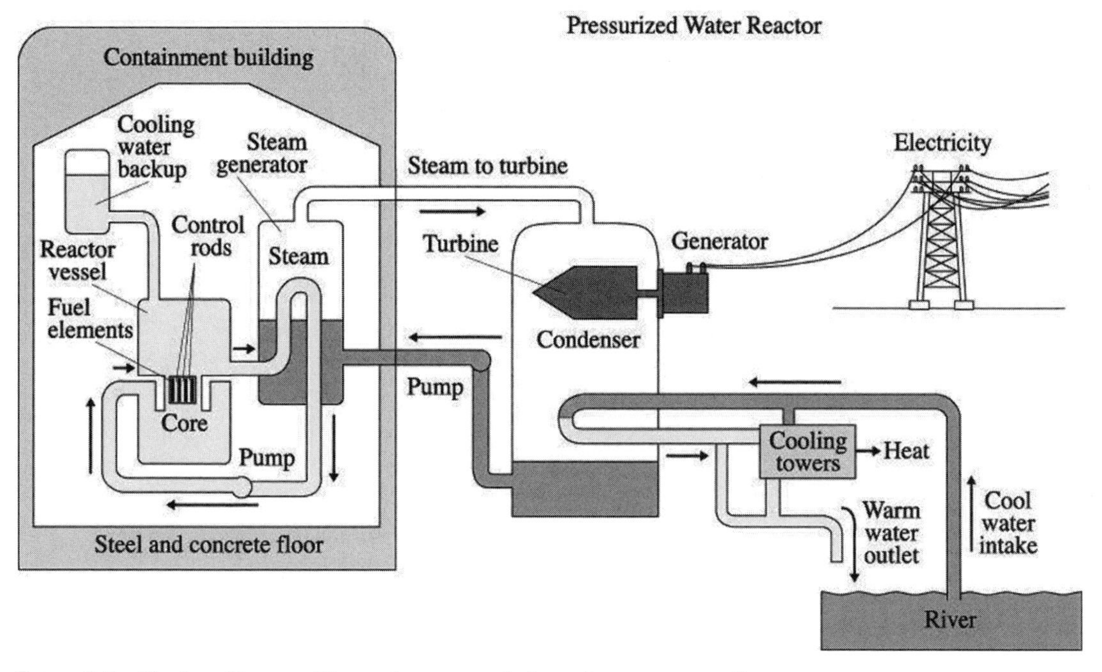

FIGURE 2-3 • Nuclear Reactor The nuclear reactor is the main component of a nuclear power plant where heat is produced by nuclear fission to convert boiling water into steam to drive turbines to generate electricity.

condensed and returned to the heat exchanger. These reactors need to be shut down each year for a period of about six weeks for refueling.

The Canadian-designed Candu, or CANada Deuterium Uranium, reactor is similar to the light water reactors but uses heavy water (deuterium oxide) as a moderator and coolant. However, unlike the light water reactors, the Candu does not need to be shut down for refueling. Besides the light water reactors, there are other types of nuclear reactors.

Boiling Water Reactor (BWR)

The second most common type of reactor is the boiling water reactor (BWR). This reactor works similarly to the light water reactor except that it has a simpler design. It uses only one section, the core of the reactor, where water is maintained at a lower pressure and is boiled to produce steam.

Fast Breeder Reactor (FBR)

A nuclear reactor that is designed to produce both power and new fuel at the same time is called a fast breeder reactor (FBR). The core of the reactor consists of bundles of plutonium-filled fuel rods surrounded by an outer layer of U-238 fuel rods. The U-238 fuel rods are bombarded by high-speed neutrons, which split apart the U-238 and causes a chain reaction in which some U-238 is transformed into plutonium-239 or

U.S. Nuclear Regulatory Commission (NRC)

An agency established by the U.S. Congress under the Energy Reorganization Act of 1974 to replace the Atomic Energy Commission (AEC), the Nuclear Regulatory Commission (NRC) ensures adequate protection of the public health and safety, the common defense and security, and the environment in the use of nuclear materials in the United States. The NRC's scope of responsibility includes regulation of commercial nuclear reactors; nonpower research, tests, and training reactors; medical, academic, and industrial uses of nuclear materials; and the transport, storage, and disposal of nuclear materials and waste. In the United States, control of nuclear-energy activities is the responsibility of the NRC. The commission grants licenses for the building and operation of nuclear reactors and for the use of nuclear materials. The construction and operation of nuclear reactors have also come under increased scrutiny by involved state and local governments.

Pu-239. Only high-speed neutrons can cause U-238 to split in a way that releases enough neutrons to continue the chain reaction and produce plutonium in sufficient amounts for extraction and processing for later use as a fuel. At the same time, the FBR generates steam that can drive a turbine to produce electricity.

The use of high-speed neutrons in breeder reactors requires special coolants other than water. Water would slow down the neutrons, preventing them from converting U-238 into Pu-239. Therefore, instead of using water to cool its core, a breeder reactor uses liquid sodium or a combination of sodium and potassium, or helium gas, substances that do not affect neutron speed.

The disadvantage of breeder reactors is that the electricity they generate is generally more costly than the electricity produced by other nuclear reactors. Breeder reactors are also expensive to construct and require a great deal more engineering to make them operational. Another disadvantage is that plutonium is much more radioactive than uranium, making disposal of FBR wastes difficult and the threat of a nuclear disaster more imposing. A small amount of plutonium can also be used to build nuclear weapons.

The average nuclear reactor has a lifespan of 20 to 40 years before it becomes obsolete and is decommissioned or retired. As an example, in 1999, a large 1,000-ton radioactive reactor was decommissioned. It was the largest U.S. nuclear power plant ever to be shut down. The power plant was shipped to the Hanford Nuclear Reservation Waste Site in eastern Washington, where it is buried 15-meters (45 feet) deep.

Pros and Cons of Using Nuclear Energy

Advantages of Nuclear Power

The advantage of nuclear power over other energy sources is its ability to create enormous energy from a small volume of fuel. One metric ton of nuclear fuel produces energy equivalent to 2 million to

3 million tons of fossil fuel. Fossil fuel systems generate hundreds of thousands of metric tons of gaseous, particulate, and solid wastes, but nuclear systems produce less than 1,000 metric tons of high- and low-level waste per plant per year. At one time, France burned coal for its electricity needs. Presently, the nation's production of electricity is 80 percent nuclear. As a result, France produces less air pollution and carbon dioxide emissions than some of its neighboring countries, who burn fossil fuels for their electricity needs.

Environmental Concerns of Nuclear Power

There are major environmental and social concerns about nuclear power, among them these:

1. The mining and the processing of uranium produces radioactive mine *tailings* which has caused contamination of land and dwellings.
2. The disposal of high-level radioactive wastes is a problem. High-level radioactive wastes include fuel or materials left after reprocessing of spent fuel from nuclear reactors and nuclear weapons production facilities. Exposure to such radiation can be hazardous to human health, and repeated exposures have a cumulative effect. It can damage reproductive cells and cause genetic mutations, producing possible physical defects in future generations.
3. Some nuclear opponents believe that existing power plants present the potential for accidents or terrorist attacks that would be disastrous because of the possible release of radiation. The 1979 accident at the Three Mile Island nuclear reactor near Harrisburg, Pennsylvania, raised questions about the safety of nuclear power. The event was given extensive media coverage, contributing to public concern about reactor safety.

International concern over the issue of reactor safety was also renewed following the Soviet Union's Chernobyl meltdown in April 1986. Chernobyl is a town in the Ukraine whose nuclear power plant suffered a catastrophic accident in 1986. The fire and explosion of the power plant created nuclear fallout and radiation that contaminated large areas of northern Europe, particularly the United Kingdom, Finland, and Sweden. Health experts expect to see a rise of cancer deaths over the next 50 years which may total as high as 40,000 cases. The explosion occurred during a test, when fuel rods became so hot that they caused a steam explosion that blew off the top of the reactor, ejecting radioactive fuel and burning control rods into the atmosphere. Risking their own lives, Chernobyl lab technicians, firefighters, and others worked under difficult conditions to control the

Spent Fuel

Once a year, approximately one-third of the nuclear fuel inside a reactor is removed and replaced with fresh fuel. The used fuel is called spent fuel. Spent fuel is the fuel rods used in a nuclear reactor that are no longer able to produce nuclear fission. Spent fuel is highly radioactive and produces considerable heat. For these reasons spent fuel must be cooled and shielded.

After the spent fuel is removed from a reactor, nuclear power plants temporarily store the used fuel in water pools at the reactor site. The water acts as a radiation shield and coolant. However, storing the spent fuel in pools is intended only as a temporary measure until a permanent disposal place is found.

In 1982, Congress passed the Nuclear Waste Policy Act, which addressed the spent fuel storage problem. The act directed the NRC to approve means of interim dry spent storage. Such storage will hold spent fuel already cooled in the spent fuel pool for at least five years. The commission authorizes nuclear power plant licensees to store spent fuel at reactor sites in NRC-approved dry storage casks. Casks can be made of metal or concrete and are either placed horizontally or stand vertically on a concrete pad above ground. The casks used in the dry storage systems are designed to resist floods, earthquakes, tornadoes, and extreme temperature.

Ten nuclear power plants in the United States are currently storing spent fuel under the dry storage option. Federal regulations permit the transportation of spent nuclear fuel only in a very strong robust metal container, the Type B transportation cask. Type B casks are designed and constructed to safely contain their radioactive contents under normal and severe accident conditions. Tests have demonstrated that this type of cask will survive the forces that it would likely experience in an earthquake, train collision and derailment, highway accident, or fire. Since 1965, there have been more than 2,500 shipments of spent nuclear fuel in the United States without injury or environmental consequences as a result of the radioactive nature of the cargo.

According to the DOE, more than 20,000 metric tons of spent fuel are stored in pools and dry casks at more than 60 nuclear power plants in over 30 states across the nation. In the year 2000, an estimated 40,000 metric tons of spent fuel were produced. Most experts believe that the need to find safe and permanent disposal of nuclear waste is becoming more and more critical because storage pools are almost full at some nuclear power plants.

TABLE 2-2 Spent Fuel Stored in Nuclear Power Stations

State	1995
Alabama	1,439 metric tons
Arizona	465
Arkansas	581
California	1,391
Colorado	15
Connecticut	1,254
Florida	1,440
Georgia	1,019
Illinois	4,292
Iowa	235

A technician is at the vault door of a waste storage area used for low-level radioactive solid wastes. The storage areas are divided into a transuranic storage area and a subsurface disposal area. Transuranic waste includes clothing, tools, and other materials contaminated with plutonium and other elements heavier than uranium. (Courtesy of U.S. Department of Energy.)

fire. Several gave their lives, and many were commended for their bravery. The Chernobyl plant was finally shut down permanently in December 2000.

As of 2001, no deaths have thus far been conclusively attributed to the operation or malfunction of any commercial nuclear power plant in the United States. Nevertheless, the potential for cancer and genetic damage as a result of the accidental release of radiation has led to increased public concern about the safe operation of reactors. Because of this concern, the development of nuclear energy for the generation of electricity in the United States has been curtailed by issues relating to safety and waste disposal. There has been some interest in using nuclear power to produce future hydrogen fuel for automobiles.

NUCLEAR FUSION

Nuclear fusion is the combining of a nuclei, such as hydrogen, into a heavier nuclei. Fusion occurs naturally within the core of the sun, where the pressure and temperature are extremely high. Commercial use of nuclear fusion as an energy source is still in the experimental stages.

Deuterium (H^2), a hydrogen radioisotope that fuels fusion reactions, is available in large amounts; however, temperatures in millions of degrees celsius are required to initiate a fusion reaction.

In the hydrogen bomb, such temperatures are provided by the detonation of a fission bomb. Sustained fusion reactions, however, require the containment of nuclear fuel at extremely high temperatures long enough to allow the reactions to take place. In 1994, U.S. researchers used deuterium and tritium (another hydrogen isotope, H^3) to achieve a one-second fusion reaction that generated about 10.7 million watts of

DO YOU KNOW?

One gram of deuterium can yield as much energy as ten tons of coal.

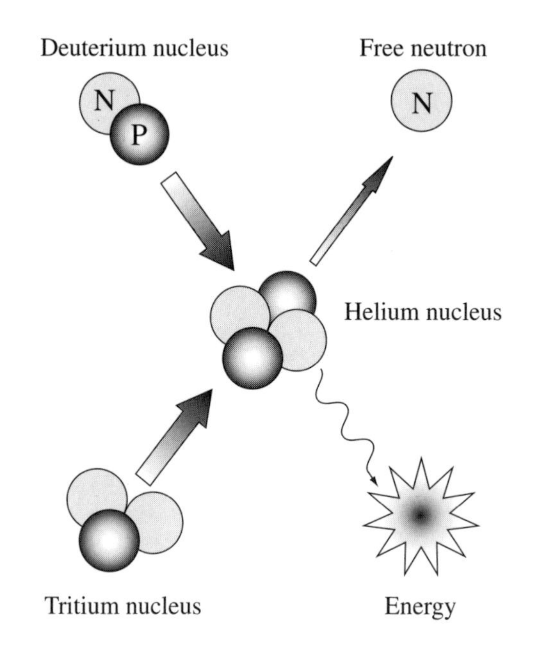

FIGURE 2-4 • Fusion
Fusion is a nuclear reaction that occurs when several nuclei combine to form a single, heavier nucleus releasing a large amount of energy.

power. Fusion reactions result in a greater conversion of mass to energy than occurs in a fission reaction. A thermonuclear explosion, for example, releases thousands of times as much energy as does an atomic bomb.

Nuclear fusion offers some advantages over the use of fissionable materials. Fusion reactors produce no fission byproducts and little radioactive wastes. And the possibility of accidents is minimized. But fusion has its drawbacks, too. Fusion reactor power plants may use toxic materials such as lithium. Illnesses and deaths can occur in humans if the material is inhaled or ingested.

THE FUTURE OF NUCLEAR ENERGY

Is there a future for nuclear energy? Most energy analysts would probably agree that a rapid growth of nuclear energy looks bleak. Throughout the world, there is public opposition for new nuclear plants, concerns about safety standards and regulations, and the potential for accidents in the present plants. Other concerns include how to remove and dispose of existing nuclear waste, as well as the challenge of decommissioning obsolete plants. There has also been a withdrawal of government *subsidies* that have assisted in the past for the construction of new plants and the operation of existing ones. However, as bleak as it may seem, the global demand for electricity is growing rapidly. If this trend continues, nuclear exponents may resurrect more creative technologies for developing new kinds of nuclear reactors and power plants to meet the growing needs for electricity. If so, the new-generation reactors will need a simpler and more rugged design, one that makes them easier to operate and less vulnerable to human and mechanical problems and the possibility of core meltdown accidents, as well as one

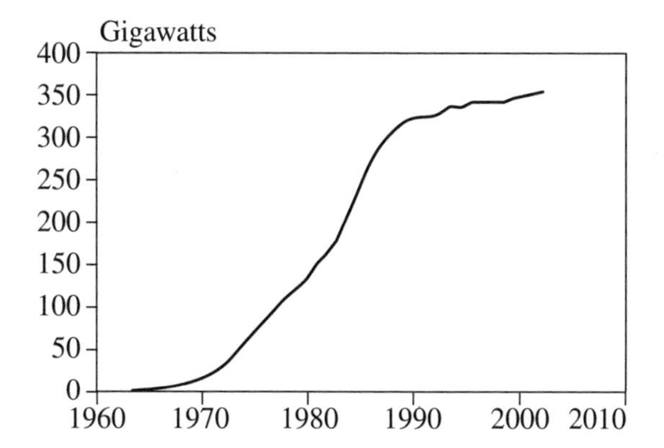

FIGURE 2-5 • World Electrical Generating Capacity of Nuclear Power Plants, 1960–2002 *Source:* World-watch Institute.

that will have a minimal effect on the environment. This new nuclear power technology may still happen as the supply of fossil fuel resources dwindles in this century. So interest in nuclear energy is still alive.

In 2003, according to the IAEA, there are several nuclear reactors under construction. India is planning to build 8 reactors, China and Ukraine will build four each, Japan and the Russian Federation will build three each, Iran and South Korea will build two each, and Argentina, North Korea, and Romania are planning to construct one each.

Vocabulary

Atoms Units of matter that consists of a nucleus and surrounding electrons

Cadmium A metallic element that is naturally present in soil and rock

Chain reaction Nuclear reaction where a neutron hits a nucleus, makes it split, and so releases further neutrons

Deuterium A chemical element that is an isotope of hydrogen containing a proton and a neutron. Heavy water contains deuterium instead of hydrogen.

Element A chemical substance, such as oxygen, carbon, or hydrogen, that cannot be broken down to a simpler substance

Isotopes Elements that have the same chemical properties as the other forms but a different atomic mass. As an example, the element carbon has several isotopes.

Nuclei Plural for *nucleus*, which is the central positively changed region of all atoms. All nuclei contain protons and neutrons with the exception of hydrogen, which has only a single proton.

Placer deposits Alluvial deposits of sand or gravel containing particles of valuable minerals

Proliferation An increase or spread at a rapid rate

Reactor core The central part of a nuclear reactor where the fuel rods are placed

Subsidies Financial assistance given by a government or other enterprise

Tailings Waste ore from mining operations

Activities for Students

1. Read the book *1,000 Paper Cranes*, which is about the aftermath in Hiroshima and Nagasaki. Then write a reflection paper on the pros and cons of the use of nuclear energy during World War II.
2. Using worldwide weather maps, determine where the fallout from the atom bombs dropped in Japan might have landed in the three years after initial impact.

3. Watch the movies *The Day After* and *Silkwood*, and then describe some of the moral and ethical challenges that the people in the Manhattan Project must have grappled with as they built the bomb. Also describe the challenges faced by the people dealing with nuclear power after the dropping of the bomb.
4. Contact your local energy companies and find out which type of reactors are used in your areas (LWR, BWR, or FBR) and what their yield is. How do the companies dispose of their waste?
5. More and more people live close to electrical and nuclear power plants. What can individuals do to ensure that they are not exposed to radioactive elements in the ground, in the water, or in the air that they breathe?

Books and Other Reading Materials

Bickel, L. *The Deadly Element: The Story of Uranium*. New York, Stein and Day, 1979.

Moss, N. *The Politics of Uranium*. New York: Universe Books, 1982.

Murray, Raymond L. *Nuclear Energy: An Introduction to the Concepts, Systems, and Applications of Nuclear Processes*. Boston: Butterworth-Heinemann, 1993.

Necker, M. *Gold, Silver, and Uranium from Seas and Oceans: The Emerging Technology*. Los Angeles: Ardor Publications, 1991.

Ramsey, Charles B., and Mohammad Modarres. *Commercial Nuclear Power: Assuring Safety for the Future*. New York: John Wiley & Sons, 1998.

Ringholz, R. *Uranium Frenzy: Boom and Bust on the Colorado Plateau*. Albuquerque: University of New Mexico Press, 1991.

Winnacker, Karl. *Nuclear Energy in Germany*. American Nuclear Society, 1979.

Wolfson, Richard. *Nuclear Choices: A Citizen's Guide to Nuclear Technology*. Rev. ed. Cambridge: MIT Press, 1993.

Websites

American Nuclear Society, http://www.ans.org

Nuclear Energy Institute, http://www.nei.org

Nuclear Information & Resource Service, http://www.nirs.org

Nuclear Information & Resource Service, www.ntp.doe.gov, www.rw.doe.gov/pages/resource/facts/transfct.htm

U.S. Department of Energy, Office of Nuclear Energy, Science & Technology, http://www.ne.doe.gov

U.S. Nuclear Regulatory Commission, http://www.nrc.gov

U.S. Nuclear Regulatory Commission, Radioactive Waste Page, http://www.nrc.gov/NRC/radwaste

Alternative Energy Resources

Alternative energy resources are those energy sources that can be used to replace fossil fuels. The principal alternative sources are wind energy, water (hydroelectric) power, solar energy, geothermal energy, biomass, and hydrogen. These alternative sources are also known as *sustainable* energy resources.

Refer to Volume V for more information on sustainable energy resources.

Although the technology for expanding the use of alternative energy sources is growing, these resources do not supply very much of

The first large-scaled Native American owned and operated wind farm is located on Rosebud Sioux reservation in south-central South Dakota. (Courtesy of Bob Gough, Intertribal COUP.)

the world's energy. As of 2000, fossil fuels supplied only about 60 percent of the world's electricity and about 80 percent of its nonelectrical energy needs.

WIND ENERGY

The fastest growing renewable power source is wind energy. Wind energy or wind power is an alternative energy resource that uses the renewable energy in moving air to generate electricity. Although wind power currently produces less than 2 percent of the world's electricity, the Worldwatch Institute estimates that wind energy could easily provide 20 to 30 percent of the electricity needed by many countries. In the United States, the American Wind Energy Association (AWEA) estimates that by the year 2025 wind power will produce more than 10 percent of the electricity in the United States.

Wind energy is the *kinetic energy* associated with the movement of atmospheric air. Wind energy systems convert this kinetic energy to more useful forms of power. Wind *turbines*, or *aerogenerators*, are used to generate electricity from wind. Most often, wind turbines are installed in large numbers called wind farms. Sites suitable for use as wind farms are usually located in areas that regularly receive sustained winds of at least 22.5 kilometers per hour (14 miles per hour) and are not be blocked by obstacles, such as high mountains.

Wind Farms and Plants

U.S. WIND FARMS

In 2001, a 300-*megawatt* wind farm was constructed on the Oregon-Washington border, and as of that date it became the world's largest wind farm. The wind farm is projected to supply electricity to 70,000 homes and businesses. But there are more windfarms on the way. A 3,000-megawatt wind farm is being built in South Dakota on the Iowa border, and when it is completed it will be 10 times the size of the Oregon-Washington

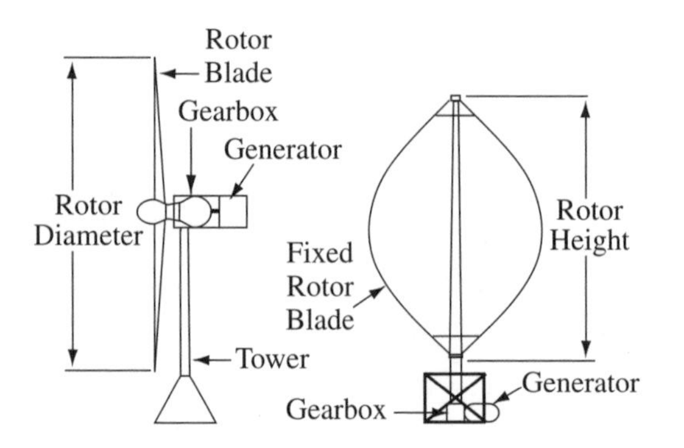

FIGURE 3-1 • Wind Turbine Designs

In the United States, wind farms are being developed in many sections of the west and mid-west. This wind farm is in California. (Courtesy of Warren Gretz/NREL, National Renewable Energy Laboratory.)

wind farm. Wind farm projects are actively being constructed in Minnesota, Iowa, Texas, Colorado, Wyoming, and Pennsylvania.

California was one of the first states to use wind farms. One of the state's wind farms, called Altamont Pass, is located east of San Francisco among a series of low hills that separate the San Francisco Bay area from the San Joaquin Valley. Another California wind farm, called Tehachapi Pass, is located in the Mojave Desert north of Los Angeles. Its 5,000 wind turbines generate enough electricity to meet the residential needs of thousands of Southern Californians, making the site one of the world's largest producer of wind-generated electricity. In all, there are about 26 states in the United States that can produce wind energy.

DID YOU KNOW?

According to the AWEA, the wind conditions in Texas, North Dakota, South Dakota, and Kansas can generate enough wind energy for the whole US.

Wind Plants of Germany

The use of wind power is growing most rapidly in Germany. In 1994, the recently reunified country surpassed Denmark as the world's second largest wind energy powerhouse. Approximately 2.5 percent of the country's electricity is produced by wind power. Wind farms in Germany are concentrated along the coastal areas in the northwest portion of the country.

Wind Plants of Denmark

The Scandinavian country of Denmark is ranked third in total wind generating capacity. The country receives about 13 percent of its electricity from wind plants. The Danes were the first people to regularly produce electricity using wind power, and by World War I they had a network of

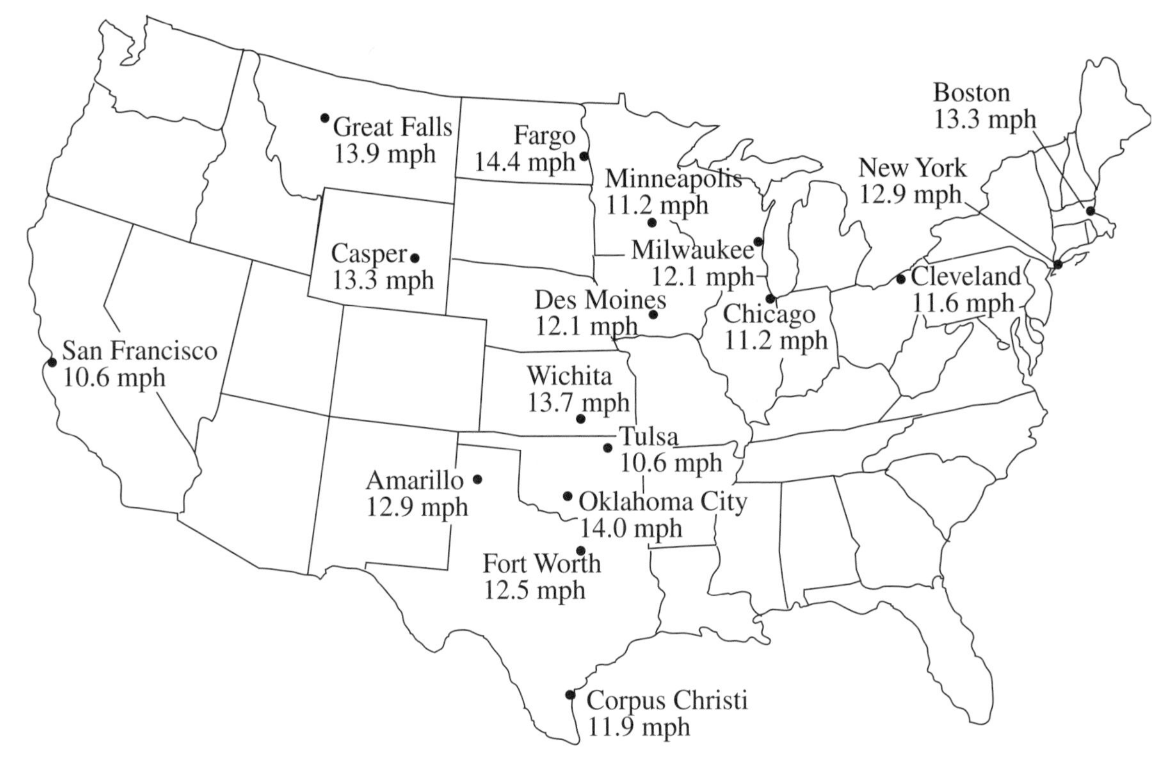

FIGURE 3-2 • The map shows wind conditions in some areas of the United States that may have the potential for wind power energy developments. California, Texas, and Minnesota are the largest wind energy states.

wind turbines that generated about 100,000 *kilowatts* of electricity. Today Denmark contains the largest concentration of wind turbines outside the United States and Germany. Approximately 75 percent of these turbines are installed in single units or in small clusters, rather than on large wind farms. More than 100,000 Danish families now own wind turbines or shares in *cooperatives* that own wind power turbines.

WIND PLANTS OF THE UNITED KINGDOM

The weather patterns and topography of the United Kingdom, particularly Great Britain, provide this region with the best potential wind resources in all of Europe. The development of wind power as a source of electricity in Great Britain began in the early 1990s, when 10 wind turbines were installed on a farm in Cornwall. Today, most of the United Kingdom's wind plants are located in England and Wales.

WIND PLANTS IN OTHER COUNTRIES

India is the second fastest growing market for wind energy. The country's 10 aerogenerators, located near Okha in the province of Gujarat, were among the first wind turbines installed in India.

The South American country of Brazil is also developing wind power technology as a major source of electricity. The goal in Brazil is to install wind turbines that will produce 1,000 megawatts of electricity by the year 2005. Other countries working on plans to increase their use of wind power include France, Italy, Greece, China, and India.

Environmental Concerns of Wind Power

Wind power is an excellent renewable sources of energy. However, the operation of aerogenerators near populations can cause noise pollution from the whirling blades. A large numbers of aerogenerators built in a field can create an unpleasant scene, and birds flying to close to them may be killed.

HYDROELECTRIC POWER

Hydroelectric power, a type of hydropower, uses the kinetic energy of flowing water to drive wheels or turbines to generate electricity. The amount of electric energy produced by the generator depends on potential energy—the pressure and the volume of the water that flows into the turbine.

Hydroelectric power accounts for about 22 percent of the world's electricity. Some of the largest hydroelectric power producers are Canada, the United States, Brazil, Norway, Russia, and China. Hydropower produces 10 to 15 percent of all electricity in the United States.

The Hoover dam, one of the largest dams in the world, spans the Colorado River and provides water and electricity for Arizona, Colorado, and Nevada. (Courtesy of U.S. Department of Interior. Bureau of Reclamation, Andrew Pernick, Photographer.)

FIGURE 3-3 • As of 2004, the Itaipú Hydroelectric Power Plant is the largest of its kind in the world. In 1995, Itaipú alone provided 25 percent of the energy supply in Brazil and 78 percent in Paraguay.

By contrast, hydroelectric power accounts for less than 2 percent of the electricity produced in the United Kingdom.

The world's largest hydroelectric plant is the Itaipú, located on the border between Brazil and Paraquay. Built from 1975 to 1991, Itaipú represents the efforts of these two neighboring countries, The Itaipú hydroelectric power plant is the largest development of its kind in operation in the world. In 1995, Itaipú alone provided 25 percent of the energy supply in Brazil and 78 percent in Paraguay.

What will probably be the largest dam in the world is now under construction in China. The Three Gorges dam project is being built on China's Yangtze (Chang Jiang) River. The construction, to be completed by 2008, will provide hydroelectric power, better navigation, and flood control.

Mini-Hydros

Building small, rather than large, hydroelectric power systems may be the trend for the future. Today small-scale hydroelectric power systems, called mini-hydro or micro-hydro systems, are being used on rivers and tributaries and in remote areas where construction is difficult. Such small-scale systems do not require the damming of rivers. These mini-hydro systems are used in China and the United States, as well as in several smaller countries, including Indonesia, Nepal, Sri Lanka, and

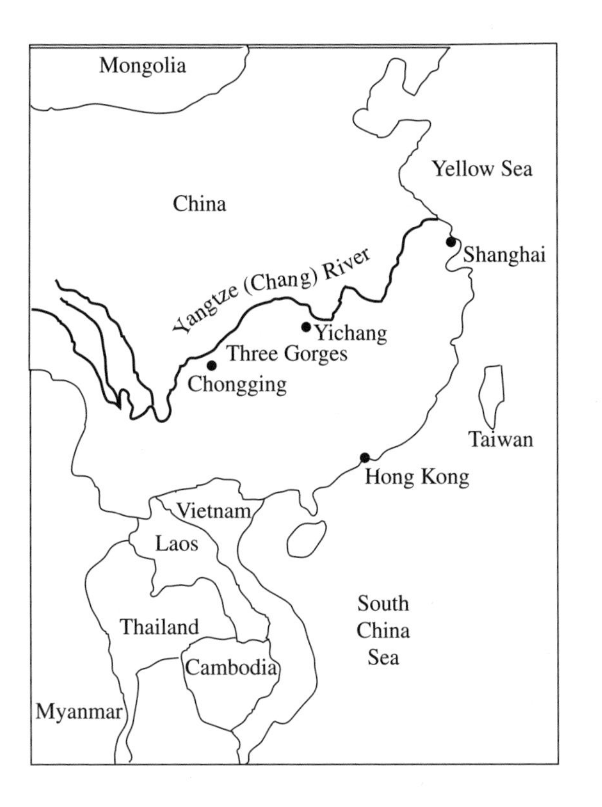

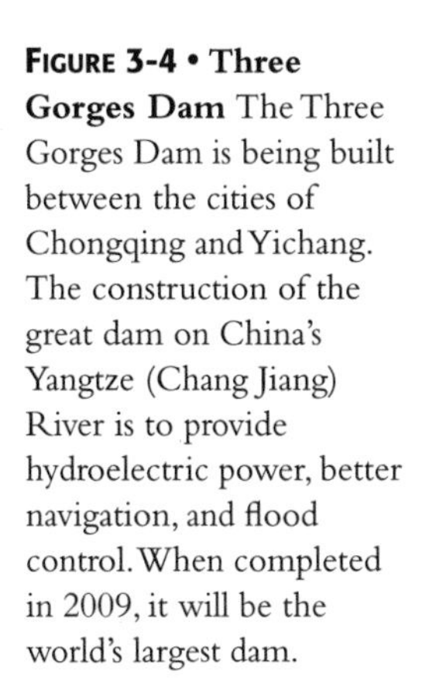
FIGURE 3-4 • Three Gorges Dam The Three Gorges Dam is being built between the cities of Chongqing and Yichang. The construction of the great dam on China's Yangtze (Chang Jiang) River is to provide hydroelectric power, better navigation, and flood control. When completed in 2009, it will be the world's largest dam.

Zaire. Small-scale hydropower stations provide the advantages of hydropower without the problems of large-scale hydropower plants. They have less impact on the environment and are less costly to build and maintain than are large hydroelectric systems. Small-scale hydropower can be used locally in remote villages and towns to generate electricity for businesses, farming needs, local lighting, and pumping water.

Environmental Concerns of Hydroelectric Power

A major benefit of hydroelectric power is that it is nonpolluting: it produces no harmful emissions such as carbon dioxide, sulfur dioxides (SO_2), and nitrogen oxides (NO_x) and no liquid or solid wastes. However, concerns about the building of large dams for hydroelectric power has led to opposition to the construction of such dams because they have negative effects on the environment. The building of a large dam requires the flooding of a large area behind the dam to form the reservoir. The flooding destroys natural habitats, wetlands, and farmlands. In addition, people living in areas designated for flooding have to be relocated. Over time reservoirs fill up with silt and sediments, burying the spawning areas of fish and other aquatic organisms. Stagnant pools can become breeding grounds for disease-causing insects and water-borne pathogens.

However, countries, such as Brazil, India, and China view the building of hydroelectric plants as one option to reduce their dependency on imported petroleum and as a preferable alternative to the use of coal-fired power plants whose emissions pollute the atmosphere.

SOLAR ENERGY

Solar energy is conversion of radiant energy from the sun into other forms of energy to provide solar heating and electricity. The two kinds of solar energy systems are passive and active. The passive solar heating system relies largely on the greenhouse effect, trapping heat inside a building much as a closed automobile traps heat when parked in an unobstructed area on a sunny day. Active systems require the use of specific materials that have been created through new technologies and also makes use of devices such as fans that can direct heat. The difference between the two systems is determined by the structures or devices each system uses to provide heat.

Passive Solar Heating Systems

The passive solar heating system relies largely on the greenhouse effect and, as noted, traps heat inside a building much as a closed automobile traps heat when parked in a shaded area on a sunny day. Passive heating requires no mechanical power or moving parts. The heating effect occurs when sunlight is absorbed by materials within the enclosed space, such as the seats in the automobile example, and the absorbed light energy is converted into heat energy, which is then radiated back into the environment.

The simplest passive solar heating system relies on how a building is positioned (the direction it faces) and the locations of its windows. Buildings designed to make use of passive solar heating are generally constructed so that their windows are placed on a south-facing side. South-facing windows are exposed to the greatest number of hours of daily sunlight throughout the year. As sunlight enters the building through the windows, light energy is absorbed by the room's walls, floors, and furnishings, which are generally composed of dark materials with rough surfaces so as to convert the greatest amount of light energy

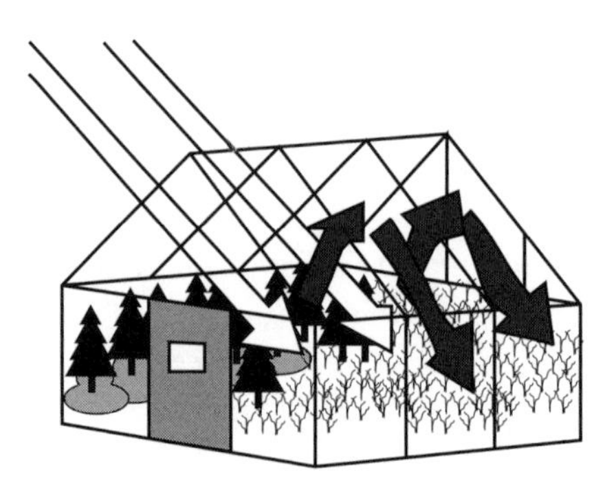

FIGURE 3-5 • The greenhouse effect describes the build up of certain gases that trap the Sun's heat in a similar way to a greenhouse. Acting like glasswalls of a greenhouse, the inside gases trap the sun's heat from escaping back into space.

into heat energy. The resulting heat energy is then slowly released back into the rooms of the structure. At night, when no sunlight is available, window shades are used to cover the windows to prevent the heat energy from escaping back into the outside environment.

Some passive heating designs feature a thermal storage wall or water-filled containers that store heat as it is generated. Vents and registers placed near these structures gather some of the heat and direct it into other rooms. Because buildings are not generally exposed to consistent amounts of sunlight throughout the year, passive solar heating systems do not completely eliminate the need for other types of heating systems that are generally fueled either directly or indirectly by fossil fuels. However, studies have indicated that, on average, the use of passive solar heating strategies in homes can reduce utility costs by as much as 20 to 30 percent.

Active Solar Heating Systems

An active solar system is more complicated than a passive system. Active systems require the use of specific materials that have been created through new technologies and also make use of devices such as fans that can direct heat. A typical simple active solar system might include three subsystems. The first subsystem is used to collect sunlight and change it to heat in a device that is usually placed on the roof of a structure that faces south. The collector, which looks like a small box, is composed of one or two panels of glass or plastic, a blackened plate, and rows of metal tubes filled with circulating gas or liquid. The metal tubes are fastened to the black plate. Sunlight entering the glass or plastic panes is absorbed and converted to heat energy by the plate. The metal tubes absorb this heat and transfer it to the fluid they contain. The heated gas or liquid is then transported to the next subsystem—a storage system.

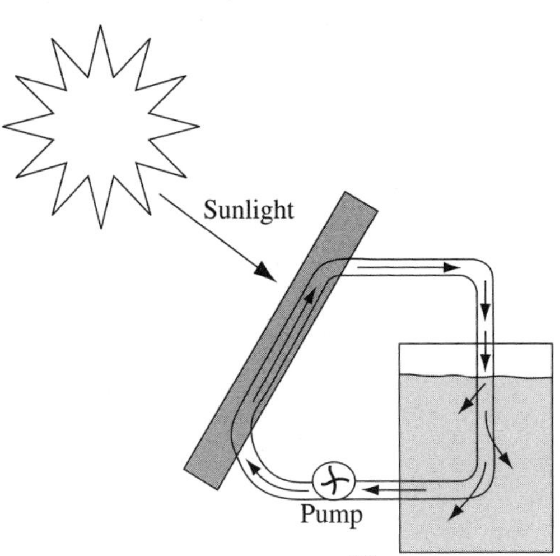

FIGURE 3-6 • Active Solar Heating System

The storage system consists of containers filled with rocks or water. Of the two materials, water works best. Water can hold heat much longer than most other materials, making it ideal for heat storage. The storage tanks are insulated to keep the heat contained inside the tank from escaping to the environment.

The last part of the subsystem is the transport system, which is composed of a network of pipes and possibly pumps or fans. The transport system circulates heated water or air from the storage system to rooms located throughout the building through a network of pipes. Pumps may be used to help move the heated water or air through the pipes. As the materials move through the pipes, they give up some of the heat they carry to the surroundings. Fans may be used to direct the released heat to needed areas. Once the heat carried by the liquid or gas is given off, the cool gas or water returns to the storage tank to be reheated.

Like a passive solar heating system, an active system may also be accompanied by a backup heating system. The backup system is used during cold periods of the year or during cloudy weather. An active solar energy system can distribute heat more effectively throughout a dwelling because of its pumps and fans. However, the pumps and fans employed by the system require electricity to operate, thus increasing the amount of energy needed to operate the system.

DID YOU KNOW?

On some South Dakota farms, solar cell–powered water pumps are particularly popular in rural areas for irrigation and other water uses. The solar PV system is cheaper than putting in power lines in these rural areas. Because of their versatility, solar cells are used as power in many rural homes and communities throughout the world.

Solar Photovoltaic Cell (PV)

A solar photovoltaic cell (PV) is a device that converts solar energy into electricity in a manner that does not release any pollutants to the environment. Today solar cells are commonly used to power small-sized items such as calculators and watches. But solar PVs have a great future in providing all the electricity needs for rural communities, homes, and businesses. In 2000, about 1 million homes were getting their electricity from home installations of solar PVs.

How Solar PVs Work

In 1839, the French physicist Edmond Becquerel observed that when light was absorbed by certain materials, the materials generated electricity. Becquerel also recorded that the amount of electricity varied with the intensity of the light. Despite these early findings by Becquerel, photovoltaic research did not begin in earnest until the 1950s. In 1952, scientists at Bell Labs in New Jersey discovered that sunlight striking a silicon-based material produced electricity. In 1958, solar PVs were first used by NASA to power the radio of the U.S. *Vanguard I* space satellite with less than one *watt* of electricity.

The process of how a solar PV works starts with sunlight. Sunlight contains energy in the form of photons, or particles of light. Therefore, the objective of a solar PV is to capture the photons of the sunlight.

Most solar PV are made from two layers of silicon that have been chemically treated using a process called doping. The doping process

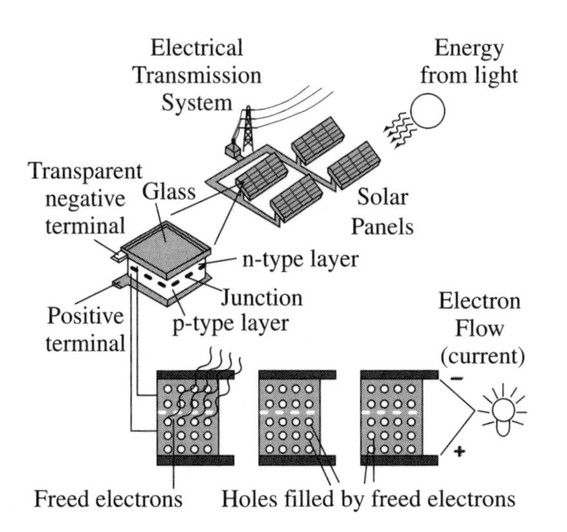

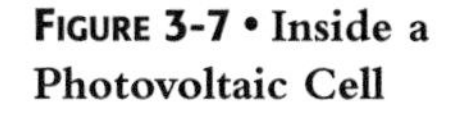
FIGURE 3-7 • Inside a Photovoltaic Cell

gives one silicon layer a negative charge (N) and the other a positive charge (P). In simple terms, the PV is simply a semiconductor wafer in which there is a junction where the N and P layers meet. Because the layers have different charges, they create an electric field at the junction that allows direct current (DC) to flow. When photons strike a PV, they cause electrons to be ejected from the silicon atoms located near the junction. The stream of electrons moves freely from the negative layer to the positive layer through the metal terminals. The metal terminals, similar to those of a battery, conduct the flow of electricity from the negative layer to the positive layer.

A 10-centimeter-square (4-inch square) solar cell can produce about one watt of direct current electricity when exposed to sunlight. One watt is not too powerful. Therefore, to generate more electricity, many solar cells are wired together in a panel called a solar array. The solar array is encased in a water-tight container. Several solar arrays, in turn, can be wired together to generate an even greater amount of electricity.

SOLAR PV MANUFACTURING

According to research, about 40 percent of all PVs sold are used for producing power for homes and for pumping water. About 35 percent of them are used in transmitting and communication operations. Solar PVs are used in lighthouses, for off-shore petroleum-drilling operations, and for use in radio and telephone transmitters. In 2000, the leading global manufacturer of solar cells was Japan, followed by the United States. In the same year, about 10 percent of all new homes built in Japan installed solar PVs. In Europe, the solar industry is lead by Germany. In South Africa, the government is planning to install 350,000 solar home systems in the country. Other countries like Italy, France, and India are also interested in manufacturing and developing PVs.

Solar Pond

The solar pond is a solar energy application that collects and stores sunlight in a body of water for the purpose of providing renewable energy. The sunlight is "captured" and stored in the bottom layer of a body of water where large quantities of salt have been dissolved. The stored salt water can attain high enough temperatures to generate electricity and to heat buildings. Solar pond technology doesn't damage the environment or produce waste materials.

Generally the salt pond has three main layers. The top layer is cold and has relatively little salt content. The bottom layer is hot, up to 100°C (212°F), and is very salty. Separating the top and bottom layers is the work of the important gradient zone. The gradient zone is between one and two meters thick (3 to 6 feet). Through a process called convection, fluids such as water and air will rise when heated up. Therefore the hot water at the bottom should rise to the surface. But the middle gradient zone stops this process. The gradient zone reduces convection. Over time, large quantities of salt are dissolved in the hot bottom layer of the body of water and become too dense to rise to the surface to cool. The gradient zone acts as a transparent insulator, permitting sunlight to be trapped in the hot bottom layer, from which useful heat may be withdrawn or stored for later use.

Solar ponds can provide a supplemental energy source for electrical production and heat for thermal desalination and space heating. Solar ponds can also provide heat for production of chemicals, foods, textiles, and other industrial products, as well as for the separation of crude oil from brine in oil recovery operations. Another use of solar ponds includes protecting fish from "cold kill" in aquaculture applications. The ponds can also function as receptacles, serving as storage containers for brine disposal from by-product wastes from crude oil production and from power plant cooling towers systems in which brine disposal is a problem.

The El Paso Solar Pond is a research, development and demonstration project operated by the University of Texas at El Paso and funded by the U.S. Bureau of Reclamation and the State of Texas. The project, which is located on the property of Bruce Foods, Inc., a food-canning company, was initiated in 1983. The El Paso Solar Pond has been in operation since 1985. It was the first in the world to deliver industrial process heat to a commercial manufacturer in 1985, the first solar pond electric power–generating facility in the United States in 1986, and the nation's first experimental solar pond–powered water-desalting facility in 1987.

SOME CHALLENGES FOR SOLAR PVs

The future for solar cell production and usage is very promising as a renewable energy resource. However, the technology still remains expensive when compared to the costs of fossil fuels. Developing countries, with limited funds, have billions of people who are not connected to electric power lines. If the costs can be more affordable, more developing countries will be in the market for purchasing solar PVs to supply electricity.

GEOTHERMAL ENERGY

Natural heat energy from Earth's interior is extracted from sources such as steam, hot water, and hot dry rocks. Known as geothermal energy, this alternative energy resource can be used for the direct heating of buildings or for generating electricity. Geothermal energy is not always listed as a renewable energy source because the depletion rate of sources such as hot water can be higher than the rate of replenishing or recharging the sources. Italy, Iceland, New Zealand, Russia, Japan, and France, along with the United States, are countries currently using geothermal energy. Other users include the Philippines, Indonesia, Mexico, countries of Central and South America, and countries in eastern Africa and in Eastern Europe.

The Geysers Geothermal Field, located in northern California, is the largest geothermal electric power plant in the world. It produces about 1,300 megawatts of electricity, enough to satisfy the residential electricity needs of 1.7 million Californians—more than the combined populations of San Francisco, Oakland, and Berkeley in the California Bay Area. According to the U.S. Energy Information Agency, geothermal energy has the potential to provide the United States with 12,000 megawatts of electricity by the year 2010, and 49,000 megawatts by 2030.

The geothermal process begins when rainwater seeps into the ground near hot igneous rocks. The groundwater is heated to form naturally occurring hot water and steam. These resources are tapped by well-drilling methods in order to generate electricity or to produce hot water for direct use.

For producing electricity, hot water is brought from the underground reservoir to the surface through production wells. The steam is separated from the liquid and fed to a turbine engine, which turns a generator. Spent geothermal fluid is injected back into parts of the reservoir to help maintain reservoir pressure. Geothermal heat sources were first used for electrical power production in Italy in 1903. The plant is located at the Larderello geothermal field, and electrical power is still being produced there.

The direct-heat energy application of hot water is used for home heating, greenhouse heating, vegetable drying, and a wide variety of

California Geothermal Geyser Field. Global geothermal geyser fields are used for generating electricity in parts of California or for the direct heating of buildings in Iceland. (National Renewable Energy Laboratory)

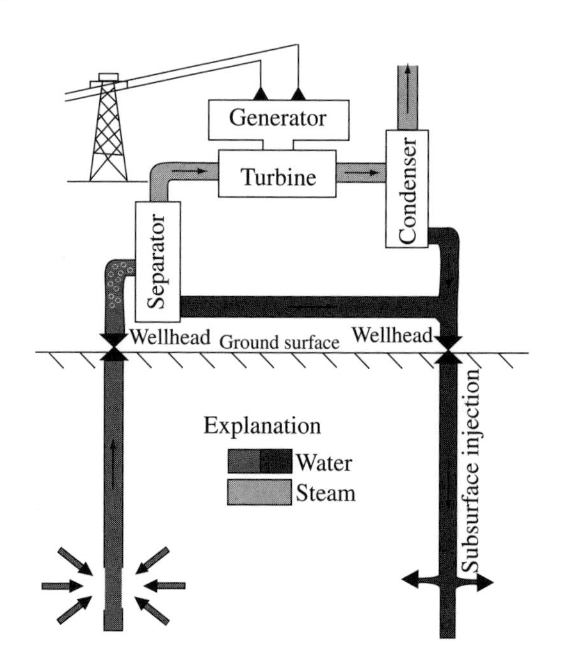

FIGURE 3-8 • A Hot Water Hydrothermal System The diagram illustrates how electricity is generated from a hot-water hydrothermal system. The part of hydrothermal water that becomes steam is separated and used to drive a turbine generator. Wastewater from the separator and condenser is injected back into the subsurface to help replace the groundwater used in the hydrothermal system. (Courtesy of Geothermal Education Office and United States Geological Survey.)

other uses. Heated water from geothermal resources can be circulated by pipes through a home or building in order to provide heat. In Iceland, for example, the entire city of Reykjavik is heated by geothermal energy.

Geothermal energy is also produced by a "hot dry rock" (HDR) system. In this technique, holes are drilled into the rocks to a depth of more than 7 kilometers (25,000 feet) where the rock temperature can exceed more than 250°C (480°F). Surface water is injected into holes or cracks where hot rocks are located. The water is heated by the rocks and rises to the surface in the form of steam, which can be used to drive a turbine to generate electricity. Many geothermal energy specialists are enthusiastic about this technique because it poses few if any environmental concerns.

Advantages of Geothermal Power Plants

Systems for use of geothermal energy have proven to be extremely reliable and flexible. Hydrothermal electric power plants operate very consistently approximately 97 percent of the time. That means that they are not shut down too often for maintenance. Nuclear and coal electric power plants average less online time performance than hydroelectric power plants. Sulfur dioxide and nitrogen oxide emissions are much lower in geothermal power plants than in fossil power plants. Nitrogen oxides combine with hydrocarbon vapors in the atmosphere to produce ground-level ozone, a gas that causes adverse health effects and crop losses as well as smog. Geothermal power plants require very little land and can be constructed quickly taking up less area than other power plants.

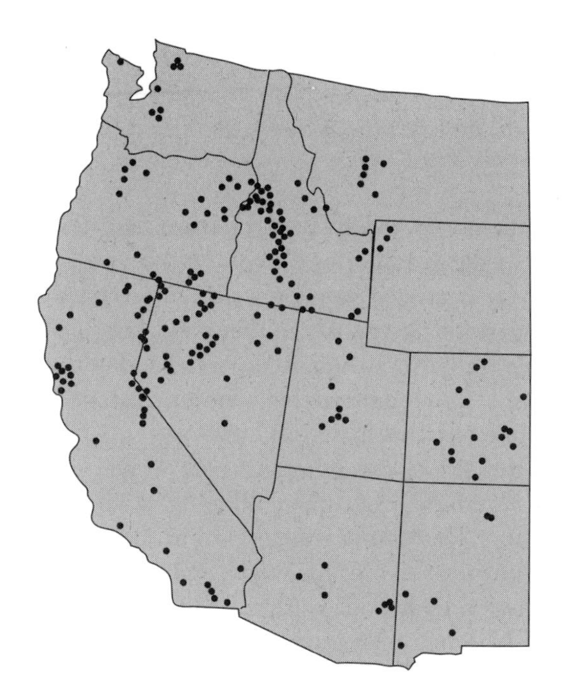

Figure 3-9 • Major U.S. Geothermal Sites

Environmental Concerns of Geothermal Energy

Geothermal energy has some disadvantages. It can be more expensive to use than other fuels. Thermal pollution can occur from the wastewater unless it is treated. The construction of geothermal plants in rainforests can destroy sensitive ecological habitats. The drilling of wells can also disrupt underground faults and fissures, which may lead to seismic activity and landslides.

The future of using geothermal energy is still cloudy. Presently, geothermal energy is not a major player in the overall energy needs of the world. However, according to advocates, geothermal energy has great potential as an energy source, but more research and development has to be initiated and financed by governments.

BIOMASS

The energy from biomass is the oldest fuel used by humans. Biomass is a term used to describe the total amount of living matter in a particular area at any given time. Fuel wood is the most widely used biomass fuel. However, there are other sources of energy from biomass. They include herbaceous plants and excess food crops that can be burned as a direct source of energy. Unused parts of sugarcane, cornstalks, peat, and even cattle dung have been used as biomass fuels. Even municipal solid wastes, a form of biomass, can also be burned directly as fuel. In Europe, processing plants use up to 50 percent of municipal trash for energy production. Energy trash–processing plants are also located in several U.S. cities in Maryland, California, Illinois, Ohio, Wisconsin, and Washington.

Fuelwood

Fuelwood is the most widely used biomass fuel. For thousands of years, wood was the world's most common source of energy utilized by humans. Today harvested wood, referred also to firewood, is used as a common source of energy for cooking and heating by as many as 3 billion people on Earth. In fact, it has only been in the last few hundred years, since the Industrial Revolution, that people have used other sources of energy, such as fossil fuels.

As late as the 1850s, wood supplied over 90 percent of the United States' energy requirements. Many developing nations still use wood as their primary fuel. Half of the energy used in the continent of Africa, for example, is in the form of fuelwood, and one-third of the world's population relies on fuelwood as a significant energy source. According to the United Nations, 1.3 billion metric tons of wood were used in 1990, either directly as fuelwood or in other energy production, representing half of all wood consumption.

Environmental Concerns of Fuelwood

There are three basic sources of wood for the production of energy: existing forests, wastes from the forest products industry, and fuelwood plantations. In much of the developing world, existing forests are the major source of fuelwood. In most cases there is no forest management, and wood may be harvested more rapidly than it can be replaced, sometimes resulting in deforestation and desertification. Dwindling fuelwood resources in many developing countries are both causing hardship to rural populations and contributing to damaging effects on the environment. Overall, the use of wood as a fuel has fewer negative impacts on the environment than the use of fossil fuels if the wood is harvested sustainably and used with properly designed stoves or other equipment.

Biofuel

Biomass contains stored energy that provides a source of fuels, known as biofuels. Biofuels are solid, liquid, or gaseous fuel derived from biomass. Biofuels are used as an alternative to fossil fuels and include biogas, biodiesel, and methane. About 5 percent of the energy consumed in the United States is provided by biofuels. Most of the biofuels are produced from wood wastes from logging operations, but they can also be produced from corn and sugar crops. In France, Italy, and Germany, biodiesel fuels are produced from

Methanol and ethanol, types of alcohol, are derived from wood, corn, and sugarcane wastes. Ethanol blended with gasoline is called gasohol. The blend consists of 10 percent of ethanol and 90 percent of gasoline.

domestic oilseeds and cottonseeds. Biofuels are cleaner than fossil fuels because they release few greenhouse gases such as carbon dioxide and sulfur and particular matter into the atmosphere.

The use of biomass fuels has the advantage over fossil fuels of being derived from renewable resources. Thus, if harvested carefully, the natural reproductive processes of living things make biomass fuels a sustainable energy resource.

DID YOU KNOW?

Some trucks and bus fleets now use biodiesel, a biodegradable, non-toxic fuel made of vegetable oils and animal fats. The mixture greatly reduces emissions and air pollutants such as hydrocarbons, sulfur, and carbon monoxide. Most diesel vehicles can run on biodiesel fuel with few or no modifications on the engines.

Environmental Concerns of Biofuels

Overharvesting of biomass for biofuels can lead to such environmental problems as deforestation, as has occurred in many parts of India, Kenya, and South American countries that use fuelwood as their primary fuel source. In addition, like fossil fuels, the burning of biomass fuels (either directly or in a gaseous or liquefied form) releases great amounts of pollutants into the environment.

OCEAN THERMAL ENERGY CONVERSION (OTEC)

Ocean Thermal Energy Conversion (OTEC) is an alternative energy resource that uses the natural temperature differences between various layers of ocean water to produce electricity. The idea of using OTEC to produce electricity is not new. A small OTEC plant was built off the coast of Cuba in the 1930s. It produced electricity for the island country until it was destroyed by a hurricane. Another plant was built in 1956 off the coast of Africa. This plant was later replaced by a dam that generates electricity as hydroelectric power at a lower cost than required to run the OTEC plant.

OTEC systems work best in the tropical waters of the central Pacific Ocean, in the Indian Ocean, and in the Gulf of Mexico region of the Atlantic Ocean. In these regions, temperature differences between warm surface waters and colder water, at depths of 1,000 meters (3,280 feet) or more, are sufficient to generate electricity.

OTEC plants can be installed on land or in the ocean. One kind of OTEC system is the closed-cycle system. The closed-cycle OTEC plant consists of pipes arranged in a closed loop. A liquid chemical with a low boiling point is placed inside the pipes. At one end of the loop is the turbine of an electric generator. Warm surface seawater at the top of the loop is pumped around the network of pipes, causing the liquid inside the pipes to be heated and changed to a gas. Movement of this gas through the pipes then causes a turbine that is connected to an electric generator to rotate and generate electricity. After passing through the turbine, the gas flows downward into the bottom part of the loop. At this stage, cold water pumped from the deep ocean is circulated around the gas-containing loop. The cold water absorbs

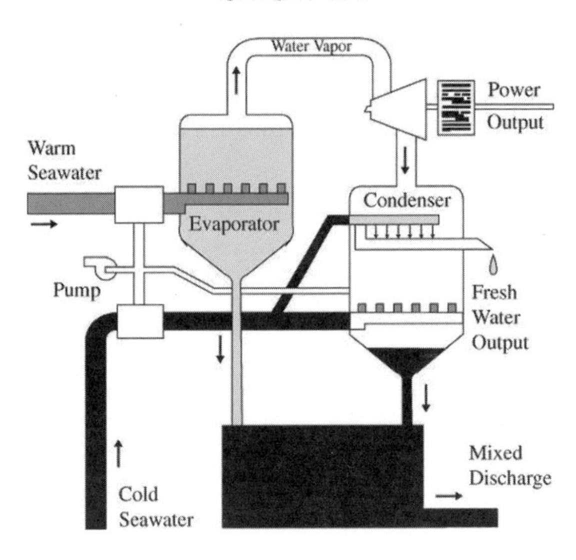

FIGURE 3-10 • Ocean thermal energy conversion (OTEC) is an alternative energy resource that uses the natural temperature differences between various layers of ocean water to produce electricity. The process begins with warm ocean water being drawn in at the top of the loop and then circulated around the pipes. The cold ocean water is used in the condenser to change the water vapor back into a liquid. *Source:* NELHA

heat energy from the loop, causing the gas to condense back into a liquid. The cycle then repeats over and over.

Another kind of OTEC system is the open-cycle system. In this type of plant, warm ocean water is boiled within a vacuum chamber. As the water evaporates, it produces low-pressure steam that is used to generate electricity. Cold ocean water is then used to condense the steam into fresh water, which can be pumped to communities for use as drinking water or to agricultural regions for use in irrigation.

OTEC research and feasibility studies are currently being conducted by the state of Hawaii and by the Japanese government. The attraction of the OTEC system is that it may be useful in generating both electricity and fresh water. Water supplied by the systems can also be used to raise seafood in commercial agriculture projects, for air conditioning, and for refrigeration. OTEC systems do have some drawbacks, however. One drawback is that the systems are useful only in tropical areas. These areas are subject to seasonal natural disasters such as hurricanes and typhoons, which can completely destroy an OTEC plant. Another drawback is that electricity produced using OTEC systems is more costly than electricity generated by methods such as hydroelectric power and the burning of fossil fuels. Unlike fossil fuels, however, OTEC systems do not release harmful pollutants into the atmosphere.

Environmental Concerns of OTEC

A major concern about the use of OTEC systems is that they may alter ocean water temperatures in the areas where they are used. Too great a change in water temperatures can affect the ability of the region to support sea life or can result in a change in the species diversity in the area.

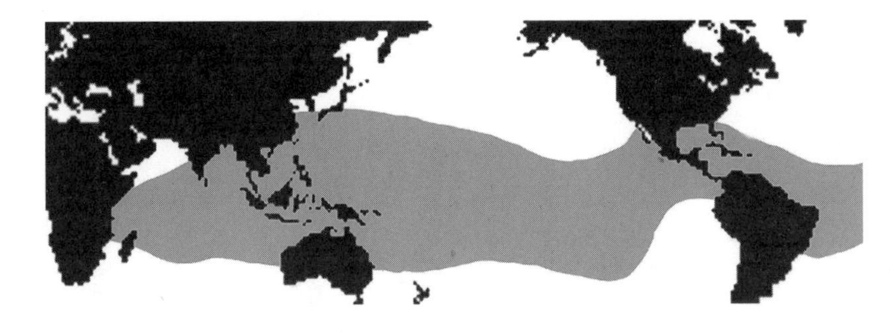

FIGURE 3-11 • Ocean thermal energy conversion systems work best in the tropical waters of central Pacific Ocean, the Indian Ocean, and in the Gulf of Mexico region of the Atlantic Ocean.

Construction of OTEC plants and laying of pipes in coastal waters may cause localized damage to reefs and near-shore marine ecosystems. Limited OTEC research continues in other countries, especially Japan, Canada, Great Britain, France, and Taiwan.

HYDROGEN FUEL CELL

Some energy consultants state that some day, a hydrogen fuel cell will be used to produce electricity to power automobiles and other machines. Hydrogen is the lightest and the most common element in the world. Today hydrogen is used primarily in ammonia manufacturing, petroleum refining, and synthesis of methanol. It's also used in NASA's space program as fuel for the space shuttles and in fuel cells that provide heat, electricity, and drinking water for astronauts. In the future, hydrogen could be used to fuel vehicles and aircraft, as well as provide power for our homes and offices. The good news is that when hydrogen is used as an energy source, it generates no emissions other than water, which is recycled to make more hydrogen.

Refer to Volume V for more information about hydrogen and fuel cells.

Hydrogen

One of the most important elements being considered as future fuel and part of a fuel cell is hydrogen. Hydrogen is an element that is a colorless and odorless gas found in water and in all organic matter, including acids. Hydrogen makes up 75 percent of the universe's mass. Its isotopes include deuterium and tritium. Hydrogen is found on Earth only in combination with other elements such as oxygen, carbon, and nitrogen. As an example, a water molecule (H_2O) contains two atoms of hydrogen and one atom of oxygen.

A fuel cell is a device that directly converts hydrogen into electricity by electrochemically combining hydrogen and oxygen without combustion. The fuel cell produces water and heat as its only waste products.

A fuel cell consists of two electrodes sandwiched around an *electrolyte*. Oxygen passes over one electrode, hydrogen over the other. This activity results in the flow of electrons known as electricity. Unlike a battery, a fuel cell does not "run down" or require recharging; it generates energy as long as fuel is supplied. In addition, fuel cells are quiet, are flexible, and operate at low temperatures.

There are many types of fuel cells. The different types of fuel cells use different electrolytes, operate at different temperatures, and are suited to different uses. For example, some fuel cells are considered best suited for use in automobiles; others are more suited for use with gas turbines.

Toyota Fuel Cell Hybrid Vehicle (FCHV). The modified sport utility vehicle powered by pure compressed hydrogen has a top speed of more than 95 miles per hour. (Courtesy of Warren Gretz/NREL, National Renewable Energy Laboratory.)

1. Hydrogen fuel is channeled through field flow plates to the anode on one side of the fuel cell, while oxygen from the air is channeled to the cathode on the other side of the cell.

Hydrogen gas
Backing Layers
Air (oxygen)
Oxygen flow field
Hydrogen flow field

2. At the anode, a platinum catalyst causes the hydrogen to split into positive hydrogen ions (protons) and negatively charged electrons.

3. The Polymer Electrolyte Membrane (PEM) allows only the positively charge ions to pass through it to the cathode. The negatively charged electrons must travel along an external circuit to the cathode, creating an electrical current.

Unused hydrogen gas
Water
Anode
Cathode
PEM

4. At the cathode, the electrons and positively charged hydrogen ions combine with oxygen to form water, which flows out of the cell.

FIGURE 3-12 • Fuel Cell Source

Source: U.S. Department of Energy

Engineers are currently experimenting with the use of fuel cells to power automobiles and buses. The vehicles are now tested in Chicago, Vancouver, and British Columbia. California has several buses and about 50 fuel cell cars up and running. Overseas, European automobile companies are experimenting with fuel-celled powered vehicles.

The fuel cell automobiles are at an earlier stage of development than are the electric vehicles (EVs) powered by batteries. However, the use of fuel cells offers several advantages over battery-powered vehicles. Among the advantages are the ability of fuel cells to be refueled quickly and to travel greater distances between refuelings. They do not need constant recharging, as do battery-powered automobiles. In addition,

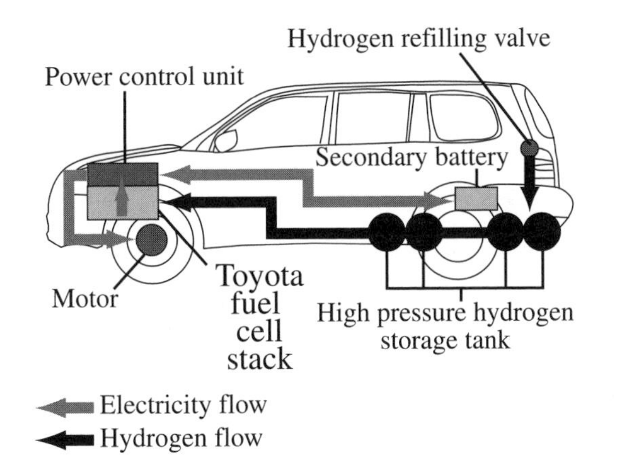

FIGURE 3-13 • FCHV4 System Diagram
Source: Toyota Motor Corporation.

automobiles powered by fuel cells are likely to produce fewer greenhouse gases than are gasoline-powered models. Although fuel cell technology is at an early stage, many scientists and environmentalists believe that switching to fuel cell vehicles can help us cut down on much of the current urban pollution from vehicle exhausts. Some experts predict that by 2020, the fuel cell will replace the vehicle combustion engine.

Some Concerns of Fuel Cells

Hydrogen researchers must overcome several obstacles if hydrogen is to become a major energy resource. Hydrogen is currently more expensive to produce than traditional energy sources, and the *infrastructure* to efficiently transport and distribute hydrogen must be developed. As an example, once fuel-celled vehicles go into production, many hydrogen refueling stations to service the fuel-celled vehicles will need to be constructed. Fuel cells may also need to be designed smaller than their present size.

Vocabulary

Aerogenerators Windmills with fast-moving sails or blades that are used to generate mechanical energy or electrical energy

Cooperative An enterprise, such as a business, that is owned and operated for the mutual benefit of people in the enterprise; also known as a co-op

Electrolyte A substance, such as sodium chloride (NaCl) that forms ions (charged particles) in a water solution, making the water an electrical conductor. For example, NaCL forms Na^+ and Cl^-

Infrastructure A basic framework of a system or organization; it can include roads, electricity, gas lines, schools, hospitals, and pipelines

Kilowatt (kW) 1,000 watts or 1.34 horsepower

Kinetic energy Energy of motion

Megawatt (mW) 1 million watts or 1,000 kilowatts

Sustainable A process that does not result in uncontrolled depletion of natural resources and has no adverse effect on the environment; a process to leave natural resources and the environment undamaged for future generations

Turbines Machines with vanes, blades, or buckets that rotate about an axis driven by the pressure of a liquid or gas. The mechanical energy produced can be used directly, or it can be converted to electrical power by transferring the turbine's torque to an electrical generator

Watt A unit of power, one joule per second; a watt is very small, it is equal to the power used to raise a glass of water from your knees to your mouth in one second. Large amounts of power are measured in kilowatts, megawatts, and gigawatts

Activities for Students

1. Why are California, Germany, and Denmark so successful with wind-powered turbines?
2. Build a solar panel to light a light bulb.
3. Draw a diagram to show how the chemical and energy dynamics of a salt pond "stores" heat in water. Why is salt the catalyst? How does this relate to OTEC in the ocean waters of Africa, Cuba, the Pacific Islands, the Indian Ocean, and the Gulf of Mexico?

Books and Other Reading Materials

American Solar Energy Society, 2400 Central Avenue, Suite G-1, Boulder, CO 80301. *Solar Today* (periodical), 54 p., $29/yr.

Berinstein, Paula. *Alternative Energy: Facts, Statistics, and Issues (Alternative Energy)*. Phoenix, Ariz.: Oryx Press, 2001.

Chandler, Gary, and Kevin Graham. *Alternative Energy Sources (Making a Better World)*. New York: Twenty-First Century Books, 1996.

Daley, Michael J. *Amazing Sun Fun Activities*. New York: McGraw-Hill Professional Publishing, 1997.

Duffield, W. A., J. H. Sass, and M. L. Sorey. (*Tapping the Earth's Natural Heat: U.S. Geological Survey Circular 1125*. Denver, Colo.: U.S. Geological Survey, 1994). A full-color book that describes, in nontechnical terms, USGS studies of geothermal resources—one of the benefits of plate tectonics—as a sustainable and relatively nonpolluting energy source. Available from the U.S. Geological Survey, Information Services Branch, P.O. Box 25286, Denver, CO 80225.

Escheverria, J. D., et al. *Rivers at Risk: The Concerned Citizen's Guide to Hydropower* (paperback). Washington, D.C.: Island Press, 1990.

Gipe, Paul. *Wind Energy Comes of Age. Wiley Series in Sustainable Design*. New York: John Wiley & Sons, 1995.

Pacific Northwest Laboratories. *Wind Energy Resource Atlas*, distributed by the American Wind Energy Association (AWEA). Washington, D.C., 1987; reprinted in 1991. Graham, Ian. *Geothermal and Bio-Energy (Energy Forever)*. Austin, TX: Raintree/Steck Vaugh, 1999.

Righter, Robert W. *Wind Energy in America: A History*. Norman: University of Oklahoma Press, 1996.

Websites

American Wind Energy Association, http://www.igc.apc.org/awea/news/html

Biodiesel Fuel, www.biodiesel.org

Center for Renewable Energy and Sustainable Technology (CREST), http://solstice.crest.org/

Energy & Geoscience Institute, http://www.egi.utah.edu

Geothermal Database USA and Worldwide, http://www.geothermal.org

International Geothermal, http://www.demon.co.uk/geosci/igahome.html

National Renewable Energy Laboratory, http:llnrelinfo.nrel.gov

Natural Energy Laboratory of Hawaii, http://bigisland.com/nelha/index.html

Solar Energy Homepage, http://www.soton.ac.uk/~solar/

Solar Energy Industries Association, http://www.seia.org/main.htm

U.S. Bureau of Reclamation Hydropower Information, http://www.usbr.gov/power/edu/edu.htm

U.S. Department of Energy, http://www.doe.gov

U.S. Department of Energy Alternative Fuels Data Center, http://www.afdc.nrel.gov

U.S. Department of Energy, Photovoltaic Program, http://www.eren.doe.gov/pv/text_frameset.html

U.S. Geological Survey, http://www.ga.usgs.gov/edu/hybiggest.html

Wind Energy Projects throughout the United States, http://www.awea.org/projects/index.html

Agencies and Organizations

American Solar Energy Society, 2400 Central Avenue, Suite G-1, Boulder, CO 80301.

Center for Environmental Resource Management, University of Texas at El Paso, El Paso, TX 79968.

Center for Renewable Energy and Sustainable Technology (CREST), Solar Energy Research and Education Foundation, 777 North Capitol St., NE, Suite 805, Washington, DC, 20002.

Land Resources: Soil and Minerals

Soil and minerals are important natural resources. Humans and other organisms depend on plants and other vegetation that grow in the soil for food, clothing, and shelter. It is important to note that only a thin layer, just a few centimeters of topsoil, supports the growth of food plants and other crops. Without this soil, there would be little or no life on Earth.

Minerals are the different substances that compose soils and rocks. In fact, minerals form the basic foundation of soil. Many rocks, such as granite, contain minerals such as quartz, feldspar, and mica. Many of the tools, jewelry, and machines we use today are produced from minerals. Minerals such as calcium, zinc, iron, and potassium are also important supplements in human diets.

Soil and minerals are nonrenewable resources—that is, they have the potential of being used by human beings faster than they are replaced by nature. It is therefore important to have a basic understanding of the properties of soils and minerals to determine their future uses and to conserve and manage them carefully.

SOIL RESOURCES

Importance of Soil

How important is soil? According to the U.S. Department of Agriculture (USDA), soil provides several important services or functions:

- Soil supports the growth and *diversity* of plants and animals by providing an environment for the exchange of water, nutrients, energy, and air.
- Soil regulates the distribution of precipitation between infiltration and runoff.
- Soil acts as a filter to protect the quality of water and other resources.
- Soil supports buildings, bridges, and other structures and protects archeological treasures.

Soil Formation

Soil makes up the outermost layer of Earth and is formed from the chemical and mechanical *weathering* of rocks. Throughout geological history, weathering has lead to soil formation. Soil formation takes hundreds or thousands of years to form.

Soil is a complex mixture made of solids, liquids, and gases. It contains particles of rocks and minerals as well as living and dead organisms. Soil composition and the rate it forms varies from one place to another. Factors such as climate, rock formation, shape and position of the landscape, and time play a major role in the formation of soil.

DID YOU KNOW?

Natural processes can take more than 500 years to form 2.5 centimeters (one inch) of topsoil.

Parent Rock

The basic material from which soil is formed is called its parent rock. Parent rocks include metamorphic, sedimentary, and igneous rock formations. The parent rock is usually the underlying rock from which the broken-up mineral particles in the soil originate. Soil that has bedrock as its parent rock is called residual soil. Here the rock is weathered *in situ*. However, sometimes Earth materials are moved and deposited away from the parent rock by the forces of gravity, by glaciers, and by wind and water. Soils formed in this way are called transported soils. Transported deposits are often moved very long distances. For example, much of the soil in New England was formed many years ago from the action of advancing and retreating of glaciers from northern areas.

Refer to Volume I to learn more about soil formation.

A typical sample of residual or transported soil contains about 40 to 45 percent minerals, 15 to 25 percent gases, 20 to 25 percent water, and 5 to 10 percent organic matter. However, these proportions can vary enormously in different geological areas.

Like other plants, lichens can wear away rocks over time. Lichens are plants that result from a symbiotic relationship between a fungus and a photosynthetic green alga or cyanbacteria. (Courtesy of Harry Cawthorn)

Some Properties of Soil

Soil Texture

The mineral particles of soil are formed from weathered rocks and are classified according to size. The different particle sizes of sand, silt, and clay, mixed in various proportions, give soil its texture. Sand, silt, and clay refer to the particle size. Sand would be the largest, whereas clay particles would be the smallest. You can feel the differences in the three particles. Sand has a gritty feel, much like that of sandpaper, and it has the thickness of a coin. Clay feels sticky, slick, and elastic, whereas silt feels more like a flour used for cooking and becomes mud when wet. The texture of soil can indicate the strength, stability, and draining of a soil. The texture of a soil also affects its capacity to retain moisture and therefore its ability to support plant growth.

Particle Sizes

Sand is a particle between 0.074 and 2 millimeters in diameter. Silt is a particle size that ranges between 0.004 and 0.074 millimeters in diameter. Smallest of all, a clay particle is below 0.004 millimeters in diameter.

Water in the Soil

Another property of water is soil. During precipitation, much of the surface water runs off the land. But some water infiltrates the ground. The water is absorbed by openings in the soil, called soil pore spaces. The water fills the tiny open spaces between the grains. The *surface tension* of a thin film of water holds the soil grains together. If water saturates the ground too much, however, there is less surface tension and the soil becomes soft and more mudlike.

Wet soils are usually the most fertile soils. These soils have high quantities of clay and organic matter. Generally the more clay (small pores) and organic matter in the soil, the larger the amount of water the soil can store. In contrast, sandlike soils are usually less fertile than clay soils because the large pores in sand allow water to drain away by gravitational flow, much like water flowing through beach sand. Therefore, in well-drained soils the amount of water available to vegetation is limited. Yet too much water in the soil can be harmful to plants and animals. Therefore, knowing the available water capacity in soils is critical when planning farming and building activities.

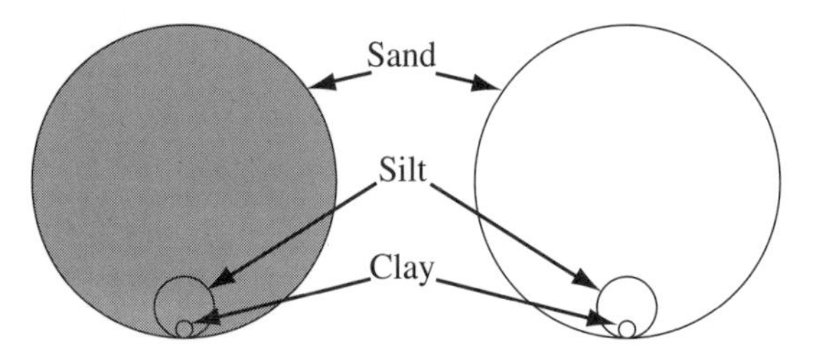

Figure 4-1 • Particle Size The relative sizes of sand, silt, and clay.

Living Organisms in Soil

The organic material in soil includes *detritus* as well as many living species. The living organisms include protozoans, fungi, insects, mites, millipedes, earthworms, slugs, moles, ants, pseudoscopions, and many others. Many of these organisms feed on the detritus. Earthworms, for example, digest organic matter, recycle nutrients, and make the topsoil richer. Fungi and bacteria break down organic matter in the soil. Mice take seeds and other plant material into underground tunnels where the plant material eventually decays and becomes part of the soil. The larger organisms continue to mix the soil while smaller organisms feed on the byproducts of the larger organisms. The cycle continues with some of the larger organisms feeding on the smaller organisms.

Organisms also play a key role in the infiltration and storage of water. Channels and aggregates formed by organisms in the soil improve the entry and storage of water. This occurs when organisms mix the mineral and organic matter as they move through the soil. The action of the organisms and the mixing activity provide pores for the movement and storage of water as well as for supplying nutrients to other organisms. The activity of the soil organism also reduces runoff and water erosion.

It has been reported that 5 to 10 tons of living organisms can live in 0.4 hectares (one acre) of soil. On a smaller scale, each gram (less than one ounce) can contain 100 million bacteria, 500,000 fungi, 100,000 algae, and 50,000 protozoa.

Gases in the Soil

Soil gases such as oxygen and carbon dioxide are also found in the pores in the soil. Roots and organisms in the soil take in oxygen and release carbon dioxide. The air in soil is usually lower in oxygen and

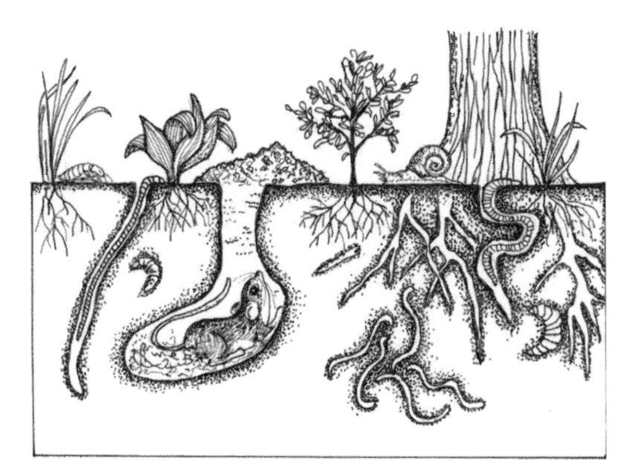

Figure 4-2 • The formation of soil begins when rocks break down and weather into smaller fragments. The process can take thousands of years. However in time, plants begin to grow in the weathered rock. Later on, fungi, algae, insects, worms, snails, and other animals add organic matter to the soil. A dark rich humus material is formed when the plants and animals in the soil die and decay. The humus provides fertile soil and helps retain water.

higher in carbon dioxide than the air in the atmosphere. So there must be a continuing exchange of air between the soil and the atmosphere. Organisms that burrow into the soil provide openings through which air and water can move.

The oxygen in the soil air is important to all life in the soil. If excess water fills the air spaces, plants and animals may die. To survive, earthworms, for example, move to the surface during a heavy rainfall when the ground becomes so waterlogged that the oxygen level in the soil drops to a deadly concentration. Plant roots cannot obtain the oxygen they need and do not grow well when there is less air in the soil.

SOIL COLOR

Color is one of the most recognizable properties that distinguish soils. Like other properties of soil, the color varies not only in different *biomes* but also at different depths. Nearly all soil colors are various mixtures of white, black, red, yellow, gray, or brown. Although black is often regarded as fertile soil, such as the soils found in the grasslands, there are exceptions such as the black soils in the tropical rainforests that are not fertile. Minerals such as hematite are responsible for red soils. Iron oxides produce green and yellow soils, and white soils can contain gypsum, kaolin, and lime materials. Grayish and bluish soil are common in poorly drained areas where soils are constantly wet.

Soil Profile

Over time the weathered materials with the parent rock are redistributed and are separated into layers called horizons. The horizons are noticeable where roads have been cut through hills. Horizons are also visible where rivers and streams have cut through their banks. The layers can be seen as a soil profile. A typical soil profile consists of several horizons.

The profile is a vertical section of a soil that is visible when a section of Earth is dug to one to two meters (three to six feet). The organic horizon is the uppermost layer called the O horizon. This layer contains fresh and decaying plants and animal remains such as leaves, twigs, needles, and other organic materials. Below this layer is the A horizon. It contains mainly mineral and organic materials and is generally darker in color than the lower horizons. The A horizon tends to be thick in savannas and thin in forest soils. The O and A layers are referred to as topsoil.

Precipitation causes leaching in the upper layers. As water moves down through the horizons, minerals and nutrients are dissolved and leached out. Materials, such as clay, are carried downward where they are accumulate in the B horizon, or subsoil. Below the B horizon is the C horizon, the parent material of weathered rocks and minerals. Underneath the C horizon lies the bedrock called the R horizon, the lowest horizon. The number of horizons and their thickness vary with different soils. Not all horizons are present in a given soil area.

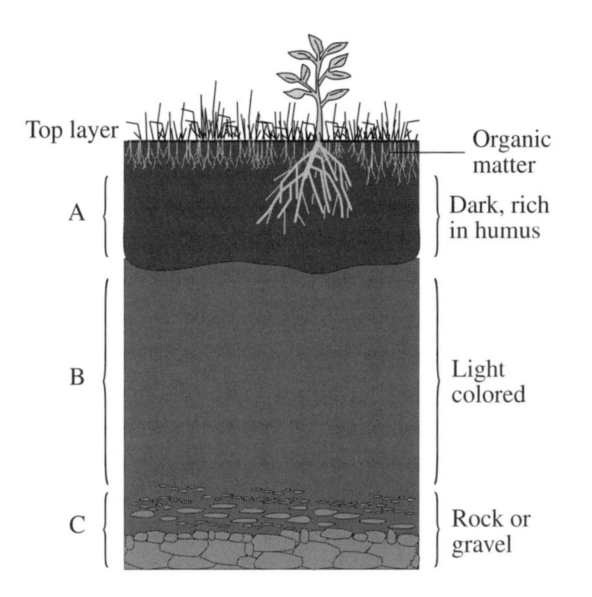

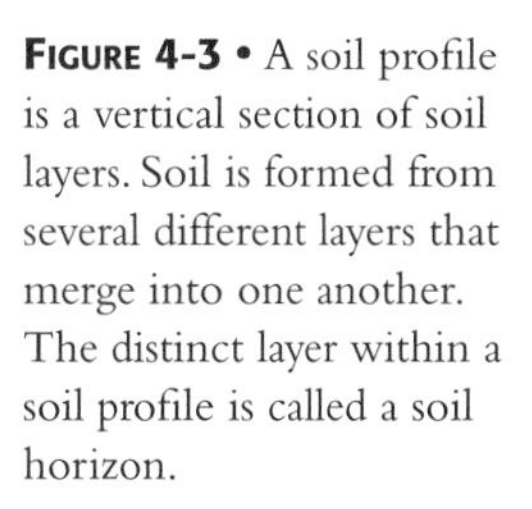
FIGURE 4-3 • A soil profile is a vertical section of soil layers. Soil is formed from several different layers that merge into one another. The distinct layer within a soil profile is called a soil horizon.

The Influence of Soil on Land Uses

Soil conditions influence how people use the land for building homes and other structures such as roads, bridges, shopping malls, recreation sites, schools, parks, and septic and sewer systems. What is put on the land is determined by the soil that is beneath it. However, no two soils are exactly the same. As an example, soils acceptable for building an airport may not be suitable for a farm. One way to help land users determine the potentials and limitations of soil is to conduct a soil survey. A soil survey generally contains soils information or data for a geographic area, usually a large section of land.

During the soil survey, soil scientists walk over the landscapes, bore holes with soil tools, and examine cross sections of soil profiles. The scientists determine the texture, color, structure, and the water content of the different soil horizons. From their research, the scientists can predict how the land can be used. They can also help the land users manage the soil resources as well. The next time you see a house, a road, or a bridge being built, you will know a soil survey was conducted first before the work began.

DID YOU KNOW?

Soil scientists of the U.S. Department of Agriculture, Soil Conservation Service (now Natural Resource Conservation Service), have identified over 70,000 kinds of soil in the United States.

World Distribution of Soils

Major Types of Soils

The major types of soils—among them tropical, desert, tundra, and temperate soils—are all based on the climates in which they form. Some climate regions have a variety of these different soil types.

Tropical Soils Tropical soils are thin and poor. Very little litter or debris is found on the surface. Much of the top horizon is made up of thin white threads of fungi and plant roots. Beneath this is usually red clay, enriched in iron and aluminum, containing few if any other minerals.

High temperatures and heavy rainfall in the tropics lead to the development of intense chemical weathering that creates infertile soils that may reach depths of three meters (nine feet) or more. The combination of weathering of rocks and high bacterial activity provides tropical soils with few nutrients and very little humus—that is, minimal organic materials. And the constant rainfall continually washes away nutrients in the topsoil.

DESERT Desert soils are produced mainly by the mechanical weathering of rocks by winds. Desert soils are light-colored, coarse, and high-salt accumulations because of the limited amount of precipitation. They contain very little humus. As a result, desert soils can support a minimum of vegetation. However, some desert soils are fertile when water is applied during irrigation.

TUNDRA OR POLAR SOILS Tundra or polar soils are found in very high latitudes and altitudes in such places as Antarctica, Canada, and Greenland. *Permafrost* appears under the thin soils. The soils are poorly drained and very shallow, usually less than a few centimeters deep.

TEMPERATE SOILS Temperate soils support a wide range of biomes that include grasslands, savannas, forests, and prairies. Grassland soils receive enough precipitation for various grasses, but not for trees. The grasslands are characterized by fertile soils. The

DID YOU KNOW?

What is your state soil? A state soil is represented by a soil that has special significance to a particular state. These official state soils share the same level of distinction as official state flowers and birds. An example is Narragansett soil, the Rhode Island state soil. The soil is a well-drained transported soil formed in glacial materials. The surface area is dark brown, containing a mixture of silt and loam. Each state in the United States has selected a state soil, 15 of which have been established by the state legislature. Soils have also been selected for Guam, Puerto Rico, and the Virgin Islands. To find out your state soil, refer to Activities for Students at the end of the chapter.

Nitrogen Cycle in Soil

In the nitrogen (N) cycle, *organic* nitrogen exists in materials formed from animal, human, and plant activities that produce manures, sewage waste, compost, and decomposing roots or leaves. These organic products transform into organic soil material called humus. *Inorganic* nitrogen comes from minerals and is added to soil from precipitation or as fertilizers. Adding N to the soil helps living plants grow and remain healthy. However, plants cannot use organic forms of nitrogen. Certain microbes living in the soil convert organic forms of N into inorganic forms that plants can then use.

Different forms of inorganic nitrogen are available to plants. Some of these may be stored in the soil (such as ammonium, NH4+). However, ammonium and other forms of nitrogen that are not held by the soil particles (such as nitrate NO_3, and nitrite NO_2) are quite soluble and can leach out of the soil and into the groundwater. Nitrogen in the form of a gas (N_2, NO, or NO_2) and ammonia gas (NH_3) escape out of the soil and into Earth's atmosphere.

There are other kinds of microbes that live in the soil and close to Earth's surface that can convert nitrogen gas into inorganic forms of N that plants can then use. When plants die, they contribute organic N to the soil. Certain microbes then convert organic N into inorganic N that living plants can use.

Scientists study water erosion in a lab to analyze runoff patterns. (Courtesy of Agricultural Research Service, USDA.)

forest soils in temperate and humid areas are not very fertile and are no more than a meter deep (three feet).

Urban Soils According to the USDA and the National Soil Survey Center (NSSC), there are soils that are characteristic urban areas. Urban soils are found in watersheds that provide drinking water, food, waste utilization, and natural resources to communities. Urban soils are also located within cities in park areas, recreation areas, community gardens, green belts, lawns, septic absorption fields, and sediment basins as well as elsewhere. Urban soils also consist of fill or disturbed soils.

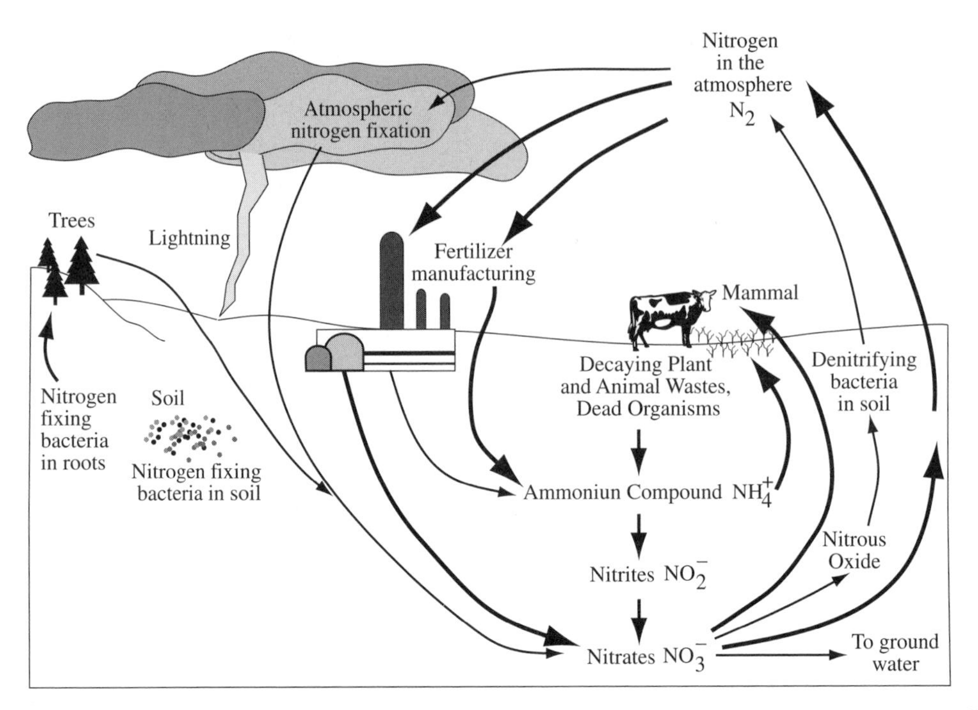

FIGURE 4-4 • This model illustrates the nitrogen cycle. It shows the conversion of nitrogen from one form to another through different processes. It is an important cycle because nitrogen is required by all living organisms.

ENVIRONMENTAL CONCERNS OF SOIL

Erosion occurs naturally by the action of chemical and mechanical weathering and the action of gravity, water, wind, and glacier action. The impact of humans on the environment has also produced soil erosion.

Refer to Volume IV to learn more about soil erosion.

According to many environmentalists, topsoil erosion is the number one serious problem throughout the world. The extensive cutting down and removal of trees in forested areas leads to soil erosion. Without the roots of trees and other plants, the land cannot hold the water and the soil is washed away. This action destroys the ability of the forest to renew itself. Overgrazing by livestock can wear away grass and other plant cover, causing desertlike conditions known as desertification. Once desertification starts, it is a major challenge to reverse the trend. Overcultivation is also responsible for topsoil erosion. Bare soil left on croplands for any period is exposed to wind and water erosion, particularly on slopes and hills. According to one estimate about 50 percent of the world's croplands are losing topsoil at a rate of 7 percent per decade.

MINERALS

In our modern society, minerals are used for jewelry, computers, cars, buildings, roads, fuels, beauty aids, and medicines. Minerals are also needed by organisms, including humans, for growth and proper functioning of body systems.

Composition of Minerals

According to pedologists, minerals form the basic foundation of soil. A mineral is a naturally occurring solid having a *crystal* structure and a definite chemical composition. Unlike coal or petroleum, which are composed of organic materials, minerals are formed through inorganic processes. In nature, most minerals exist as *compounds*, such as sodium chloride. But a small number of minerals are found in the *element* form, such as silver.

The majority of minerals are originally formed when once-heated Earth material magma (molten rock) cools and forms solid igneous rock. During the cooling process of magma, *ions* become bonded together through to electrical attraction. The attracted, bonded ions remain fixed in position and produce solid crystalline minerals within igneous rock. In this manner the crust of Earth was formed and continues to form.

Earth's crust contains a combination of naturally occurring elements, of which eight elements are most common: oxygen (O), silicon (Si), aluminum (Al), iron (Fe), calcium (Ca), sodium (Na), potassium (K), and magnesium (Mg). Combinations of these elements along with the other naturally occurring elements that form Earth's crust produce a wide variety of minerals.

DID YOU KNOW?

Minerals that contain silicon and oxygen are called silicates (SiO_4). They are the commonly occurring minerals because silicon (Si) is the most abundant element in Earth's crust. Two major silicates are feldspar and quartz. Silicates are currently very important to the computer and electronics industries. The silicates are the largest and the most complicated class of minerals by far. Approximately 30 percent of all minerals are silicates, and some geologists estimate that 90 percent of Earth's crust is made up of silicates. Silica sand is a form of quartz (SiO_2).

Mineral Groups

Mineral resources can be divided into two major groups, metallic and nonmetallic. Metallic minerals include of gold, iron, copper, lead, zinc, tin, and manganese. Many of these metals are derived from *ores*. Some of the nonmetallic minerals include halite, quartz, bauxite, borax, asbestos, talc, and feldspar. Building and ornamental stones include slate, marble, limestone, traprock, travertine, and granite. It would be impossible to cover all the mineral resources in this chapter, so the text will deal with only a selection of the most common ones based on their abundance and use.

DID YOU KNOW?

Mica is a mineral that is widely used in eye shadows, powder, lipstick, and nail polish. Mica is added to give luster or a pearl-like look to a product. One of the first ingredients listed in eye shadow is usually a magnesium silicate mineral.

METALLIC MINERALS

IRON One of the most abundant of Earth's metals is iron. It has been used as far back as 3000 B.C. and has been used to produce cast iron, wrought iron, and steel. Iron is also an important nutrient in the human diet. Iron helps to carry oxygen from the lungs to the rest of the body. Some examples of the minerals in iron ore are hematite and magnetite. The world's largest mine is located in Brazil. In the United States, the largest resource of iron is located in Minnesota's Mesabi Range.

ALUMINUM Like iron, aluminum is also an abundant metal in Earth's crust. It is mined from bauxite ore and tropical soils, much of it in the tropical forests. This lightweight corrosion-resistant, malleable metal is used in cooking utensils, foil wrapping, beverage and food containers, electrical appliances, and building materials. It is used in paints and fireworks, and it is also used to produce glass, rubber, and ceramics. It is used in several chemical forms, including aluminum nitrate, aluminum oxide, aluminum hydroxide (used in antacids), and aluminum chlorohydrate (used in deodorants). Aluminum sulfate, or alum, is used in wastewater treatment plants as a *coagulant* to remove particles.

Environmental Concerns of Aluminum Illness in humans and pollution of the environment are central concerns. People who breathe in high levels of aluminum particles in air may have respiratory problems, including coughing and asthma. Excessive levels of ingesting aluminum particles in the air can cause a lung disorder called aluminosis. Recycling of aluminum cans has reduced the number of cans ending up in landfills. About 50 percent of all aluminum cans are melted down and recycled. Aluminum production also uses a great deal of electricity.

COPPER Copper (Cu) is a metal element that is an important mineral in the human diet. Many multivitamin pills contain copper. Copper, like other metals, is soluble in water and can be absorbed into the body.

Copper is used principally to make electric wire and other electrical products because it is an excellent conductor of electricity, easy to bend, and very resistant to corrosion. Copper is also used in marine paints and wood preservatives. Much smelting is necessary to separate the copper from other metals such as zinc and lead. The smelter stacks discharge sulfur dioxide as well as lead and copper. Smelting operations can contaminate water resources and adversely affect habitats and organisms.

Environmental Concerns of Copper Copper from mine tailings can leach into the soil and cause contamination of surface water and ground water resources. Very high levels can cause adverse effects such as damage to the liver. Copper is also toxic to aquatic organisms.

LEAD Lead (Pb) is a soft, heavy, tasteless, naturally occurring heavy metal found in small amounts in all parts of the environment, including air, food, water, and soil. Lead has been used in paints, glazes, and enamels since about 1000 B.C. The ancient Romans used lead to build water pipes. Today lead is used in the manufacturing of gasoline, paints, plumbing supplies, roofing supplies, pesticides, and batteries. Lead is also used in radiation shields for protection against X-rays, as well as in the manufacturing of surgical equipment and computer circuit boards. The commonest sources of exposure to lead pollution comes from

TABLE 4-1 Minerals Used for Domestic Products

Product	Minerals
Carpet	Calcium carbonate, limestone
Glass/Ceramics	Silica sand, limestone, talc, lithium, borates, soda ash, feldspar
Linoleum	Calcium carbonate, clay, wollastonite
Glossy paper	Kaolin clay, limestone, sodium sulfate, lime, soda ash, titanium dioxide
Cake/Bread	Gypsum, phosphates, aluminum
Plant fertilizers	Potash, phosphates, nitrogen, sulfur
Toothpaste	Calcium carbonate, limestone, sodium carbonate, fluorine
Lipstick	Calcium carbonate, talc
Baby powder	Talc
Counter tops	Titanium dioxide, calcium carbonate, aluminum hydrate
Household cleaners	Silica, pumice, diatomite, feldspar, limestone
Jewelry	Precious and semiprecious stones
Kitty litter	Diatomite, pumice, volcanic ash
Potting soil	Vermiculite, perlite, gypsum, zeolites, peat
Paint	Titanium dioxide, kaolin clays, calcium carbonate, mica, talc, silica, wollastonite
Concrete	Limestone, gypsum, iron oxide, clay
Wallboard	Gypsum, clay, perlite, vermiculite, aluminum hydrate, borates
Pencil	Graphite, clay
Ink	Calcium carbonate
Sports equipment	Graphite, fiberglass
Pots and pans	Aluminum, iron
Medicines	Calcium carbonate, magnesium, dolomite, kaolin, barium, iodine, sulfur, lithium
Television	35 different minerals
Automobile	15 different minerals
Telephone	42 different minerals

Source: Women in Mining Education Foundation.

industrial activities such as lead-smelting operations and battery recycling plants.

Environmental Concerns of Lead Compounds of lead are hazardous substances and air pollutants. Exposure to lead occurs mostly from breathing workplace dust, eating contaminated foods, and ingesting lead in house paints and house dusts, especially by children who are most vulnerable to the effects of lead. The U.S. Department of Housing and Urban Development reports that lead hazards exist in 500,000 homes occupied by young children. The lead found in painted walls of older homes can be very harmful to young children, because the paints may contain very large amounts of lead. The paint in these houses often chips off and mixes with dust and dirt. Some old paint, when dry, contains 5 to 40 percent lead. Exposure to lead can be dangerous for unborn children and also for preschool children who put many things into their mouths and are more sensitive to lead's effects. Lead is a toxic metal in the soil.

MERCURY A metallic chemical element that occurs naturally in the environment as part of several stable compounds, mercury (Hg) is most commonly found in the mineral cinnabar, which is a mercury sulfide (HgS). In its elemental form, mercury is a shiny, silver-white, odorless liquid. It can combine with other elements, such as chlorine, carbon, sulfur, or oxygen, to form mercury compounds.

Mercury and its compounds have many uses. They are often used to make thermometers, barometers, batteries, lamps, skin care products, and medicinal products. Mercury compounds are also used to make neon lights, fungicides, paints, plastics, electrical equipment such as switches, and paper and paper goods. Mercury is, in addition, a component of dental fillings.

Environmental Concerns of Mercury Because of its high *toxicity*, mercury is regulated by the Environmental Protection Agency (EPA) under the Safe Drinking Water Act and the Clean Air Act (CAA).

ZINC Zinc is used in dry cell batteries, wood preservatives, disinfectants, and galvanized steel, as well as in a variety of alloys because of its resistance to corrosion. Zinc is a bluish white metallic element whose principal ores include sphalerite and calamine smithsonite that are converted to the metal. It is also a dietary supplement.

Refer to Chapter 2 for more information about uranium.

URANIUM Uranium exists naturally combined with oxygen in minerals such as pitchblende and carnotite. Uranium minerals are widely distributed in Earth's crust. They are present in sandstones, in veins within rock fractures, and in placer deposits—that is, ore materials that have been transported and deposited in river deltas and streams. Most uranium mined in the United States derives from sandstone deposits. Worldwide, the richest deposits of uranium are present in France, the Russian Federation, Ukraine, Australia, Canada, and southern Africa.

MAGNESIUM The most common sources of magnesium include dolomite, magnesite, and natural brine and salt water. Magnesium is the lightest commercially used metal in the manufacturing of aircraft parts, lightweight machinery, photographic flash bulbs, and portable tools. Magnesium is used as a neutralizer of stomach acids and as a laxative.

Manganese A silver-colored metallic element, manganese (Mn) is present in many types of rocks. Small amounts of manganese are essential for the lives of humans and other organisms. Although manganese can react to form a number of compounds, it does not break down in the environment. Low levels of manganese are present in lakes, ponds, streams, and particles suspended in air.

In industry, manganese is combined with iron to make steel and other alloys. Manganese compounds are also used to make a variety of products, including pesticides, fertilizers, ceramics, paints, dry cells, disinfectants, antiseptics, and a dietary supplement that helps the body use vitamin B1.

Heavy Metals

A heavy metal is a metallic element that is a hazardous substance and a toxic chemical, even at low concentrations. Heavy metals can cause adverse health conditions in humans and other organisms because they tend to accumulate in the food chain. They remain in the environment because they cannot be broken down. Among the common heavy metals are mercury, chromium, lead, arsenic, copper, zinc, and cadmium.

Heavy metals are widespread in the environment and can be found as particulate matter in the atmosphere and in waterways or dissolved in soil. Heavy metals are released into the environment by the combustion of leaded gasoline; in emissions from manufacturing plants, foundries, incinerators, and smelting operations; and through leaching from landfills. Most of the contamination of soil comes from heavy-metal wastes from sewage sludge, wastewater, and ash.

Environmental Concerns of Manganese In high amounts, most manganese compounds can be highly toxic. In humans, high levels may be carcinogenic and teratogenic, or harmful to the developing fetus.

Refer to Volume IV for more information on the toxicity of heavy metals.

Valuable Minerals and Gems The two principal precious metals are gold and silver. Both of these metals have been valued and used for thousands of years because of their rarity, purity, and inertness. They are valuable ornaments and a symbol of wealth. Gold occurs naturally as a free element and is used for a large variety of jewelry. The largest use of silver compounds is in photography. Sterling silver is used widely for tableware and silver coins. Other valuable gems include rubies, diamonds, and amethyst.

DID YOU KNOW?

Gold has been used historically as a colorant. Ancient Egyptians used gold to color skin and hair. Gold can still be found in powders and makeup to add a "rich" golden sheen to the skin.

Nonmetallic Minerals

Phosphorus Phosphorus (P) is a natural chemical element that is essential for every form of life on Earth. Phosphorus forms the basis of a very large number of chemical compounds, the most important of which are the phosphates. Phosphates play an essential role for living organisms. Phosphates are necessary for the functions of metabolism, photosynthesis, and nerve and muscle action; they are also important in the formation of deoxyribonucleic acid (DNA) and other cell components. Human bone is composed largely of phosphorus in the form of calcium phosphate.

Almost three-fourths of the total phosphorus in all its forms used in the United States goes into the production of fertilizers. Other important uses for phosphorus include nutrient supplements for cattle and other animal feeds, water softeners, food additives, coating agents for metals, and some pesticides.

Much of the phosphorus on Earth is tied up in sedimentary deposits and a number of phosphate-bearing rocks. The largest phosphate mine in the world is located near Tampa, Florida. Another important source is on the west coast of South America. The mine provides about 30 percent of the world's production of phosphate. Another source of phosphorus is guano (bird droppings), which builds up in places where there are large communities of nesting sea birds. The guano dries up into a rocklike mass that can be mined and processed.

Environmental Concerns of Phosphorus Scientists are concerned that several activities are helping to add phosphorus to the environment, altering its natural balance. Some of the excess phosphorus comes from sewage treatment plants and some is the result of runoff from cattle feedlots. Most, however, comes from the commercial fertilizers that wash from farms into rivers and lakes. Excess phosphorus is known to result in poor water quality and the eutrophication of lakes and rivers. Eutrophication is the process by which lakes and other water sources become so rich in nutrients as to cause a change of the kinds of organisms found in the lake. To reduce levels of phosphorus in the environment, scientists recommend the use of natural fertilizers as opposed to chemical fertilizers that contain phosphorus. Phosphorus is also a diminishing natural resource.

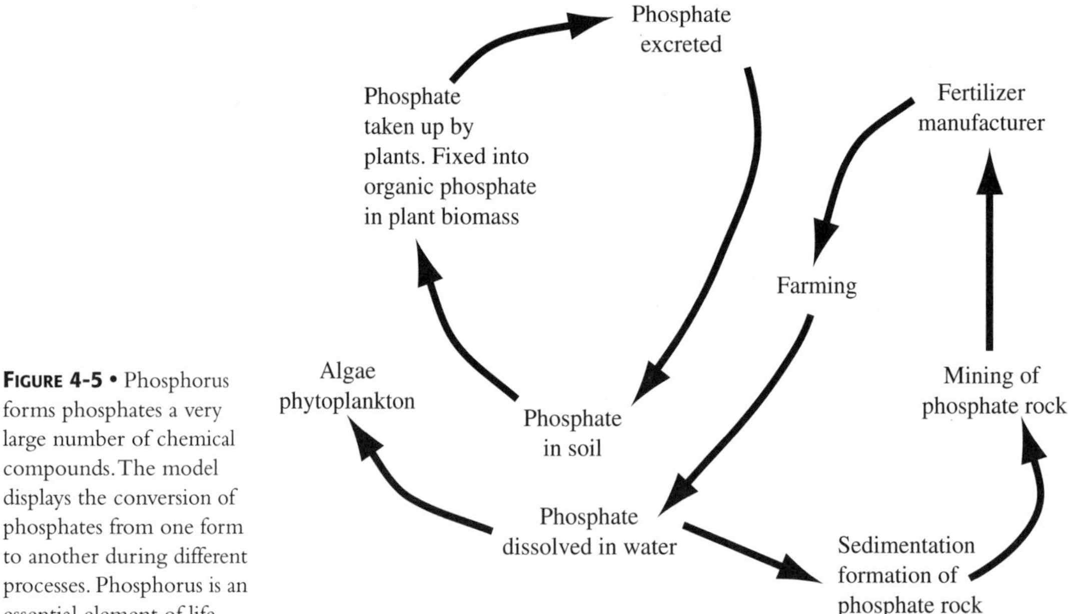

FIGURE 4-5 • Phosphorus forms phosphates a very large number of chemical compounds. The model displays the conversion of phosphates from one form to another during different processes. Phosphorus is an essential element of life.

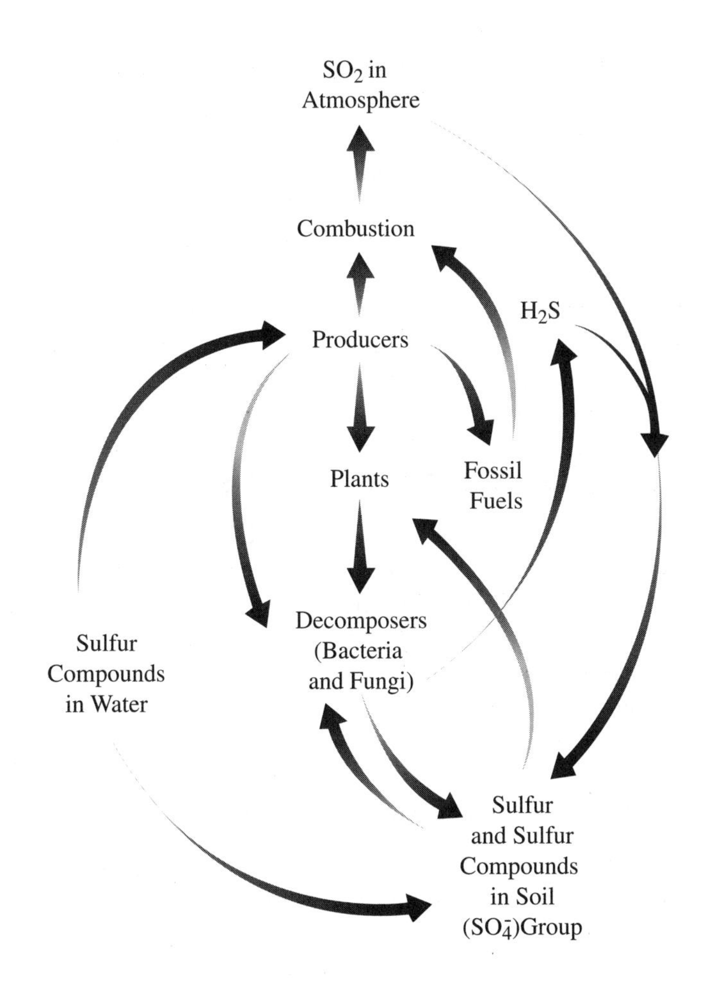

FIGURE 4-6 • This model illustrates the sulfur cycle. Sulfur is found in all fossil fuels and combined with metals. Sulfur forms several gases including sulfur dioxide, a major air pollutant, and hydrogen sulfide.

SULFUR Sulfur (S) is a yellow, odorless, nonmetallic element that occurs in elemental form or combine with metals such as copper, zinc, and mercury. It is found in all fossil fuels. Sulfur burns with a blue flame and is used in the manufacturing of sulfuric acid, gunpowder, and fertilizers. Sulfur oxides are formed when one sulfur atom combines with different numbers of oxygen atoms. Sulfur dioxide is a sulfur compound that contains one sulfur atom and two oxygen atoms.

Environmental Concern of Sulfur Sulfur oxides are a major source of air pollution. Sulfates are released when coal is burned. They contribute to acid rain and if inhaled can cause respiratory illnesses. At high levels, sulfur dioxide can cause respiratory and cardiovascular diseases.

ASBESTOS A family of fibrous, silicate minerals, asbestos was widely used in the manufacture of a variety of products worldwide until it was discovered to be a health hazard to humans. Because of its unhealthful effects, asbestos use is now banned in the United States and many European countries. Its use nonetheless continues in many developing nations throughout the world.

Halite Table salt (sodium chloride) is one of many salts. It is produced from the mineral halite. Sodium chloride is necessary for life processes. Early human communities traveled long distances to obtain salt when it was not found locally. Halite, along with other minerals, forms when ocean water is allowed to dry at the shore. Halite is also formed in underground deposits that were once ancient salt lakes.

Building Materials and Ornaments Quarried rocks such as granite, limestone, marble, and sandstone are used as building stones for homes, walls, statutes, and monuments. Crushed rock and gravel are used in road building.

Mining and Smelting the Minerals

Mining is the process of obtaining useful minerals from Earth's crust. The process includes excavations in underground mines and at the surface. Most minerals are found in large ore deposits and in veins and seams in rocks. Gold, diamonds, tin, and platinum are often found in placers, or deposits of sand and gravel containing particles of the mineral. The heating process that is used to extract metals from ore is called smelting.

Environmental Concerns of Mining and Smelting

Refer to Volume IV to learn more about the environmental concerns of mining and smelting.

Worldwide use of minerals concerns to environmental scientists because most minerals are being used at a faster rate than they can be replaced by nature. In addition, the excavation of minerals often harms the environment by destroying habitat, making land more vulnerable to soil erosion and landslides, or by degrading water quality. Smelting releases sulfur dioxide emissions into the atmosphere.These air pollutants can contaminate the soil and damage forests as well.

Vocabulary

Biomes A large region consisting of a group of ecosystems that have similar climate conditions and plant life

Coagulant A substance that makes suspended particles in water clump together

Compounds Gases, liquids, or solids that are made up of two or more different kinds of atoms or elements and bonded together

Crystal The distinctive shape (molecular arrangement) formed by an element or compound when it becomes a solid

Detritus Dead organic matter such as leaves, twigs, and the remains of other organisms that exist in the ecosystem

Diversity The rich variety and number of species in an area

Element A gas, liquid, or solid that is made up of one and only one kind of atom or element

Inorganic A substance that does not come from a living organism

In situ In place

Ion An atom, a group of atoms, or a compound that is electrically charged when the loss or gain of electrons occurs; ions include anions, which are negatively charged, and cations, which are positively charged

Ore Minerals or a combination of minerals from which a useful substance, such as a metal, can be extracted and marketed at a price that will yield a profit

Organic Refers principally to materials derived from organisms composed chiefly of C, H, O, and N elements; organisms that are alive or were alive

Permafrost Ground in which soil is permanently frozen

Surface tension The properties of a liquid surface that form a thin elasticlike film; surface tension causes all free liquids to take a spherical shape unless other forces are present

Toxicity The potency of harmful or poisonous material on a living organism when exposed to the material in various ways

Weathering The breakdown of rock into smaller pieces by chemical, physical, or biological actions

Activities for Students

1. Gather four or five samples of soil from your local area. Examine their color, water percentage, degree of erosion, and mineral content.
2. How does the harvesting of wood affect the butterflies in the Brazilian rainforest?
3. Aluminosis is caused by the inhalation of aluminum particles. Is there a correlation between the levels of aluminum particles near aluminum mines and the number of lung specialists in the hospitals near the mines?

Books and Other Reading Materials

The Alliance to End Childhood Lead Poisoning (202-543-1147). *Lead in Your Home: A Parent's Reference Guide.* EPA Booklet 747-B-98-002, 1998. Comprehensive new guide to educate parents and homeowners about lead hazards and lead-poisoning prevention in the home.

Chesterman, Charles Wesley. *National Audubon Society Field Guide to North American Rocks and Minerals.* New York: Random House, 1979.

National Science Teachers Association. *Dig In! Hands-On Soil Investigations.* The National Science Teachers Association, 1840 Wilson Boulevard, Arlington, VA 22201-3000.

Necker, M. *Gold, Silver, and Uranium from Seas and Oceans: The Emerging Technology.* Los Angeles: Ardor Publishing Company, 1991.

Prinz, Martin. *Simon and Schuster's Guide to Rocks and Minerals.* New York: Simon & Schuster, 1978.

Singer, Michael J., and Donald N. Munns. *Soils: An Introduction.* New York: Prentice Hall, 2001.

Sparks, Donald L. *Environmental Soil Chemistry.* San Diego, Calif.: Academic Press, 1995.

Websites

EPA, http://www.epa.gov/lead/

Mineral Resources, http://minerals.er.usgs.gov/

Soil Science Society of America, http://www.soils.org/sssagloss/search.html

U.S. Department of Agriculture for Kids, http://www.usda.gov/news/usdakids/index.html

USDA-NRCS National Soil Survey Center (for information on your state soil), http://www.statlab.iastate.edu/soils/photogal/statesoils/list1.htm

Women in Mining, http://www.womeninmining.org/

Land Resources: Forests

Refer to Volume I, to learn more about forests.

Forests are ecosystems dominated by trees, with a unique combination of plants, animals, microorganisms, soils, and climate. The trees in the forests are critical to the welfare of our planet. They help keep the air clean by filtering pollutants and reducing the risk of global warming by absorbing carbon dioxide and other greenhouse gases.

The majority of Earth's species are dependent on the survival of trees. As *watersheds*, forests absorb rainfall and slowly release it into streams and rivers, moderating both floods and droughts and regulating water flows. Trees are also a renewable land resource that provides the world's fuelwood, lumber and timber, paper products, food, and medicines.

Areas of Temperate Forest

FIGURE 5-1 • Areas of Temperate Forests in the Eastern and Western Hemispheres

FORESTS' CONTRIBUTIONS

For thousands of years, early civilizations were dependent on trees for such purposes as homes, rafts, canoes, fuel, and weapons. Berries, fruits, nuts, and even the bark of some trees were important foods for humans and animals. The Adirondack Indians of northern New York State used the bark of a number of conifers for food. Bark was such an important part of their diet that their name in the Mohawk Indian language means "tree eaters."

Leaves of palms and other trees were used for thatching roofs and making baskets. Cloth and woven fabrics used for clothing were made from bark, leaves, and other tree parts. Wood utensils were fashioned from coconuts and other fruits. Medicines were obtained from trees, as were dyes, tanning materials, and spices. The Quinault tribe of northwestern Washington and the Karuk tribe of northern California peeled, dried, and boiled the roots of the Pacific yew to make tea. The Quinaults drank the liquid as lung medication, whereas the Karuks used it to relieve stomach aches and kidney problems.

DID YOU KNOW?

In a cave near Beijing, China, human remains have been unearthed that are about 4,000 years old, indicating that wood was used then for either heating or cooking or both. Wood was also used to build the throne from the tomb of Tutankhamen (King Tut), circa 1350 B.C. The throne is made of cedarwood veneer overlaid with ebony and ivory.

Fuelwood

Even today people depend on forests for their medicine, clothing, timber, food, and even fuelwood. Fuelwood, a biomass fuel, referred to also as firewood, is used as a common source of energy for cooking and heating in many countries.

For thousands of years, fuelwood was the world's most common source of energy utilized by humans. In fact, only in the last few hundred

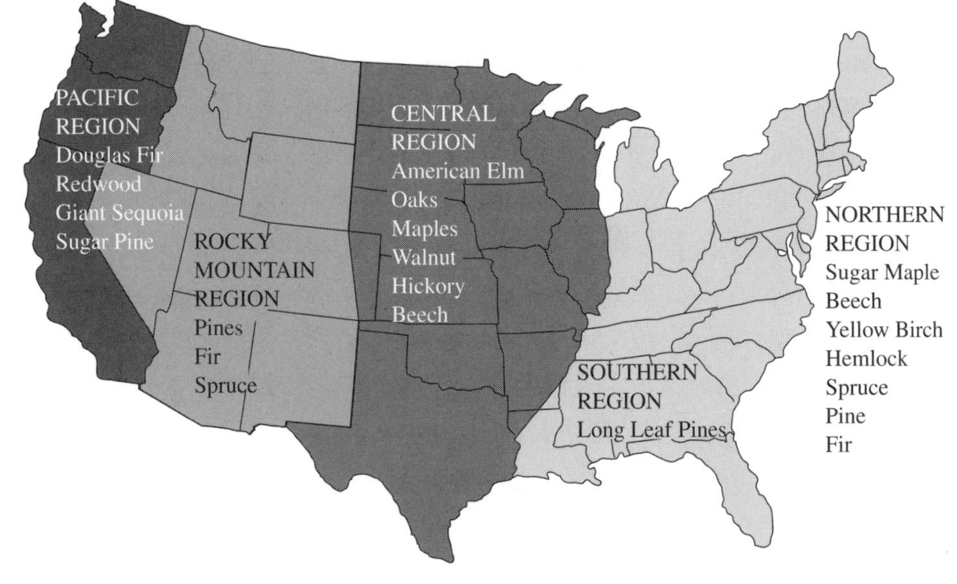

FIGURE 5-2 • Major U.S. Forest Regions

years, since the Industrial Revolution, have people used other sources of energy, such as coal, petroleum, and natural gas.

As late as the 1850s, wood supplied over 90 percent of U.S. energy requirements. Many *developing nations* still use wood as their primary fuel. Africa gets about 50 percent of its energy needs from fuelwood. About 15 percent of the energy use in Latin America comes from fuelwood, and Asian countries uses fuelwood for about 11 percent of their energy needs. In all, one-third of the world's population still relies on fuelwood as a significant energy source.

According to the United Nations, 1.9 billion cubic metric tons of wood is used annually either as fuelwood or in other energy production, representing half of all wood consumption. More than 50 percent of the world's fuelwood is produced in Brazil, Indonesia, China, India, and Nigeria. There are three basic sources of wood for the production of energy: existing forests, wastes from the forest products industry, and fuelwood plantations or tree farms.

Environmental Concerns of Fuelwood

In much of the developing world, existing forests are the major source of fuelwood. However, in most cases there is no forest management. As a result, wood may be harvested more rapidly than it can be replaced, sometimes resulting in excessive cutting down of trees, known as *deforestation*. Dwindling fuelwood resources in many developing countries are causing hardship to rural populations and contributing to damaging effects on the environment.

Overall, the use of wood as a fuel has fewer negative impacts on the environment than the use of fossil fuels if the fuelwood is harvested sustainably and used with properly designed stoves or other equipment. The use of fuelwoods and other biomass fuels has the advantage over fossil fuels of being derived from renewable resources. Thus, if biomass fuels are managed and harvested carefully, they remain a sustainable energy resource as a result of the reproductive processes of living things.

TABLE 5-1 Fuelwood Countries

Africa	Asia	Latin America
Botswana	Afghanistan	Haiti
Cape Verde	China	Peru
Ethiopia	India	Bolivia
Kenya	Nepal	
Mali	Pakistan	
Rwanda		
Sudan		

Agroforestry and Tree Farms

Agroforestry

Agroforestry techniques that include growing tree species for the purpose of providing fuelwood can help in countries where deforestation is a problem. Agroforestry is the practice of cultivating trees in combination with food crops and sometimes livestock. The growth of crops and trees together benefits both types of plants. Crops benefit from the moisture given off by the trees through transpiration; trees benefit from the nutrients returned to the soil, a process that occurs after the crops have been harvested and their wastes are permitted to decompose naturally. Trees are planted to provide shelter for crops such as coffee or to support vine crops such as vanilla and pepper. Agroforestry also benefits the environment by preserving natural forests and their habitats and helps to reduce the soil erosion that often occurs when forests are cleared to create land for agriculture.

Tree Farm

A tree farm or tree plantation is an area of land on which trees are grown and managed for commercial uses, such as fuelwood, pulpwood for paper uses, Christmas trees, and ornamental decorations. Presently, most tree farms are set up in temperate climates. However, more and more tree farms are being established in tropical regions, where trees can grow much faster than in cooler climates. There are several teak farm plantations in Southeast Asia, for example.

Often a tree farm consists of only one or a few species of trees. In some areas, tree farming is conducted to maintain a sustainable yield of wood or wood products. For example, after most of the white pine forests in the state of Maine disappeared, for example several large paper companies began tree farming to provide the pulpwood needed for their products. One of the earliest examples of tree farming occurred in the late 1800s when British scientist Henry Wickham collected more than 70,000 seeds from the rubber plant *Hevea brasiliensis* in the Amazon jungle for the purpose of growing trees in Kew Gardens, London, England. Later the saplings were transported to Ceylon, India, and Malaya, where they were used to establish the rubber industry.

Wood Products

Wood has become an important part of everyday life. According to the American Forest and Paper Association, wood is energy efficient. For example, it takes more energy to produce steel frames than to produce wooden ones. Five times more energy is used in manufacturing aluminum siding than in constructing wood siding. Steel, both new and recycled, uses 4,000 times more coal, oil, and gas in its refining, manufacturing, and fabricating process than does wood. Wood is also a good insulator. It has eight times the thermal resistance or insulating capability of concrete, 413 times that of steel, and 2,000 times that of aluminum.

Today lumber, interior trim and molding, siding, shakes, wall paneling, beams, kitchen cabinets, and flooring are just some of the wood products that are used to frame and build houses and other structures. In fact, more than a million homes are built with wood in the United States every year. Approximately 40 percent of the lumber used to build homes in the United States is manufactured in the Pacific

DID YOU KNOW?

The famous forests of cedars of Lebanon were virtually eliminated in lumbering operations during early historic times. Much of the wood was used for the construction of King Solomon's great temple and palace.

TABLE 5-2 Uses of Wood

Alder	fencing stakes
Ash	golf club shafts, axe handles, billiard cues, ladders
Beech	furniture, carpenter's tools, mallets and handles
Birch	modern Swedish furniture, clothes pegs, bowls, spoons
Blackthorn	walking sticks and chimney sweep brooms
Chestnut	often used in fencing and poles for scaffolding
Elm	keels and rudders in boat building
Hazel	baskets, golf club shafts, pegs
Larch	boat planking
Maple	bowls
Oak	furniture, deckchairs, charcoal
Pine	telegraph poles, fences, picture frames
Poplar	baskets
Sycamore	plates and food utensils
Walnut	elaborate carving in furniture
Willow	baskets, furniture

Northwest. Other products made from wood include automobile components, furniture, sporting goods, toys, and musical instruments.

DID YOU KNOW?

The ancient Egyptians, more than 5,000 years ago, made a kind of paper from papyrus, a reedlike plant.

DID YOU KNOW?

The first paper merchant in America was Benjamin Franklin, who helped to start 18 paper mills in Virginia and surrounding areas.

Paper Products

Many people think of paper as having always been a product made from wood. But the history of paper made from wood began only in the mid-nineteenth century. What we would today consider paper was first made some 2,000 years ago in China from hemp and other materials. In fact, the Chinese-manufactured paper is very similar to the kind we use today.

Paper is used to make a variety of products—everything from tea bags and coffee filters to paper tissues and beverage containers. Approximately 1.5 million tons of construction products are made each year of paper, including insulation, gypsum wallboard, roofing paper, flooring, padding, and sound-absorbing materials.

Making Paper

Most paper in general use today is made by extracting a fibrous material called *cellulose*. Cellulose is obtained from deciduous trees or evergreen (coniferous) trees. The trees are ground up and mixed with water to make a mash mixture called pulp. The papermaker can add wastepaper at this stage to make a finished recycled paper. The pulp is treated with chlorine gas to bleach the paper white. The addition of chlorine helps prevent the yellowing of the paper with age. Dyes and other materials are also added to the pulp. For the best grade of paper, more energy and chemicals are needed in the manufacturing process. The final pulp mixture is then sent to a paper-making machine, where heated cylinders press and dry the pulp into paper.

Approximately 33 percent of the world's wood harvest is used to manufacture paper. Only about 25 percent of used paper is currently being recycled. Most used paper is thrown away to end up in landfills. The United States, Japan, and Canada are the world's leading producers of paper. The United States is the leading user of paper.

DID YOU KNOW?

Paper wasps and hornets also make a paperlike nest from plants and wood materials. The insects chew these materials into a pulplike structures to make their nests.

TABLE 5-3 Paper Products

Paper Products	Examples
Containerboard	Corrugated (cardboard) boxes, box dividers, reusable shipping pallets, cushioning material
Cotton Fiber Papers	Fine stationery, paper money, maps, onionskin
Kraft Packaging Papers	Grocery bags, yard waste bags, shopping bags, pet food bags, brown wrapping paper
Newsprint	Newspapers, advertising flyers
Paperboard	Milk and beverage cartons, cereal boxes, paper plates and cups, shoeboxes, board for binder covers
Printing-Writing Papers	Books, magazines, copy paper, fine stationery, catalogs, direct mail pieces, envelopes, business forms, school filler paper
Specialty Papers	Coffee filters, automotive filters, sandpaper, greaseproof paper, parchment paper
Tissue	Bathroom tissues, facial tissue, towels, napkins

TABLE 5-4 Top 10 Consumers in 1995 of Paper and Paperboard per Person

1.	United States
2.	Finland
3.	Belgium
4.	Japan
5.	Canada
6.	Singapore
7.	Taiwan
8.	Switzerland
9.	Denmark
10.	New Zealand

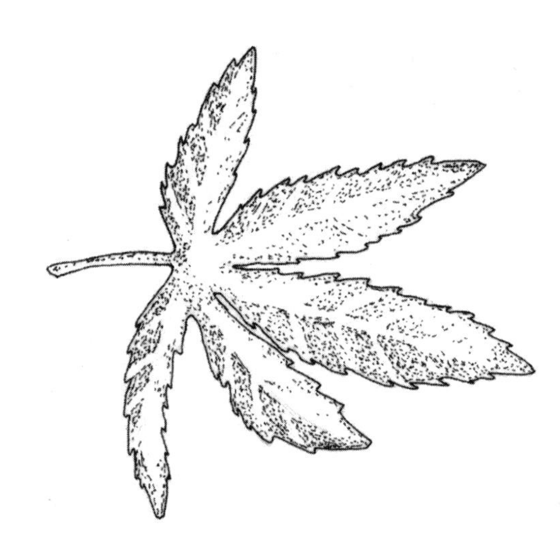

FIGURE 5-3 • More than 1,000 middle schools in Japan grow, study, and make paper from the plant, kenaf, each year. Kenaf is a cane-like plant that grows in moist southern climates such as Georgia, Texas, and Mississippi. Many experts believe that kenaf is an ideal fiber crop to replace or substitute for wood pulp used in paper making.

Bagasse: A Potential Pulpwood Substitute

Recent years have seen increased use of paper made, not from wood pulp, but from materials like bagasse. Bagasse is the fiber remaining after juice is extracted from sugarcane and from kenaf, an Asian plant similar to jute, an East Indian plant. For several years now, these materials have been used by some companies to make business cards. Bagasse paper products were used in all copier and printer paper at the global warming conference held in Kyoto in December 1997. Bagasse is obtainable without cutting down trees. Its use is expected to grow as its cost continues to drop and its quality increases.

Non-Timber Products: Food and Other Materials

Plants in the forest holds secrets for safer pesticides for farmers. Two species of tropical potatoes have sticky material on their leaves that attracts, traps, and kills predatory insects. This self-defense mechanism by these two potato plants may lead to further research to reduce the need for using pesticides on potatoes.

Trees are sources of other products besides timber, fuelwood, and paper. In the *temperate* zone, edible fruits produced by trees include apples, cherries, peaches, and pears. Maple syrup is processed from maple trees in the temperate and colder regions. Avocados, figs, persimmons, and citrus fruits can be found growing on trees in warm-temperate and subtropical regions. Important fruits in tropical regions include breadfruit, coconuts, and mangoes. Desert regions include date trees and nuts. The coconut, the oil palm, and the olive, all sources of oils and fats, are used as food and for other purposes. Trees also provide spices that include cinnamon, cloves, and nutmeg. Chocolate, coffee, and kola nuts are harvested from trees.

Nonedible tree products include fibers, resin, waxes, rubber, turpentine, creosote, and cork. One important *chemical* extracted from trees is to be found in tannins. Tannins are substances that are derived primarily from the bark of trees. One source of tannins is from the

Forest Management

Forest management is the controlled and regulated harvesting of trees, called seed-tree cuttings. Seed-tree cutting is a type of selection harvesting or selection cutting in which mature, economically desirable trees are left standing in an area that has been cleared of other trees. The mature trees, called seed trees, are left uncut as a source of seeds to regenerate the area. Seed-tree cutting is a sustainable forestry practice designed to replenish tree populations in forest areas where trees are harvested as a source of timber.

Historically, the goals and objectives of forest management have reflected the changing needs and values associated with forests. Traditional objectives revolved around timber production. By contrast, many present-day forest managers apply the principles of multiple use and sustained yields. They also set up plans to conserve the biodiversity of the forest and to preserve recreational values.

Other forest management objectives include special attention to the presence of endangered species or sensitive areas, such as steep slopes and in and around water habitats. Ease of transportation also affects how a forest is managed; the cost of road building and maintenance influences which practices can be carried out and where. Ownership and the size of the forest in question are additional factors.

eastern hemlock tree that is widely distributed across eastern North America. During the nineteenth and early twentieth century, the tannins were used to tan sheepskins and heavy leather for shoes in the United States. Today, the tanning industry in the United States relies on tannin imports or uses chemical substitutes for tanning processes. Tannins are also used in the production of dyes and inks, and as an ingredient in many beverages, such as cocoa, tea, and red wines.

Medicinal Uses of Forest Products

Forest trees and plants are sources for many medicines, herbs, and other pharmaceutical products. Taxol is a plant-derived anticancer drug that was extracted from the bark of the Pacific yew. The yew is a small- to medium-sized tree that grows in the understory of *old growth forests* in the northwest region of Canada and the United States. Taxol can also be found in the needles and bark of other related species. Preliminary studies and clinical tests by the National Cancer Institute suggest that taxol is active against ovarian cancer in some patients. This is welcome news, for approximately 12,500 women die each year from ovarian cancer. Taxol also has the potential for treating other cancers, including breast cancer.

To learn more about old growth forests, refer to Volume I.

Of all of the forest types, the tropical rainforest is the most abundant source of medicinal plants that are used in today's medicines. One-fourth of the medicines available are derived from plants in the tropical rainforests. Chemicals from tropical plants are used in surgery and for internal medicine. They are used to treat illnesses ranging from headaches to contagious diseases such as malaria.

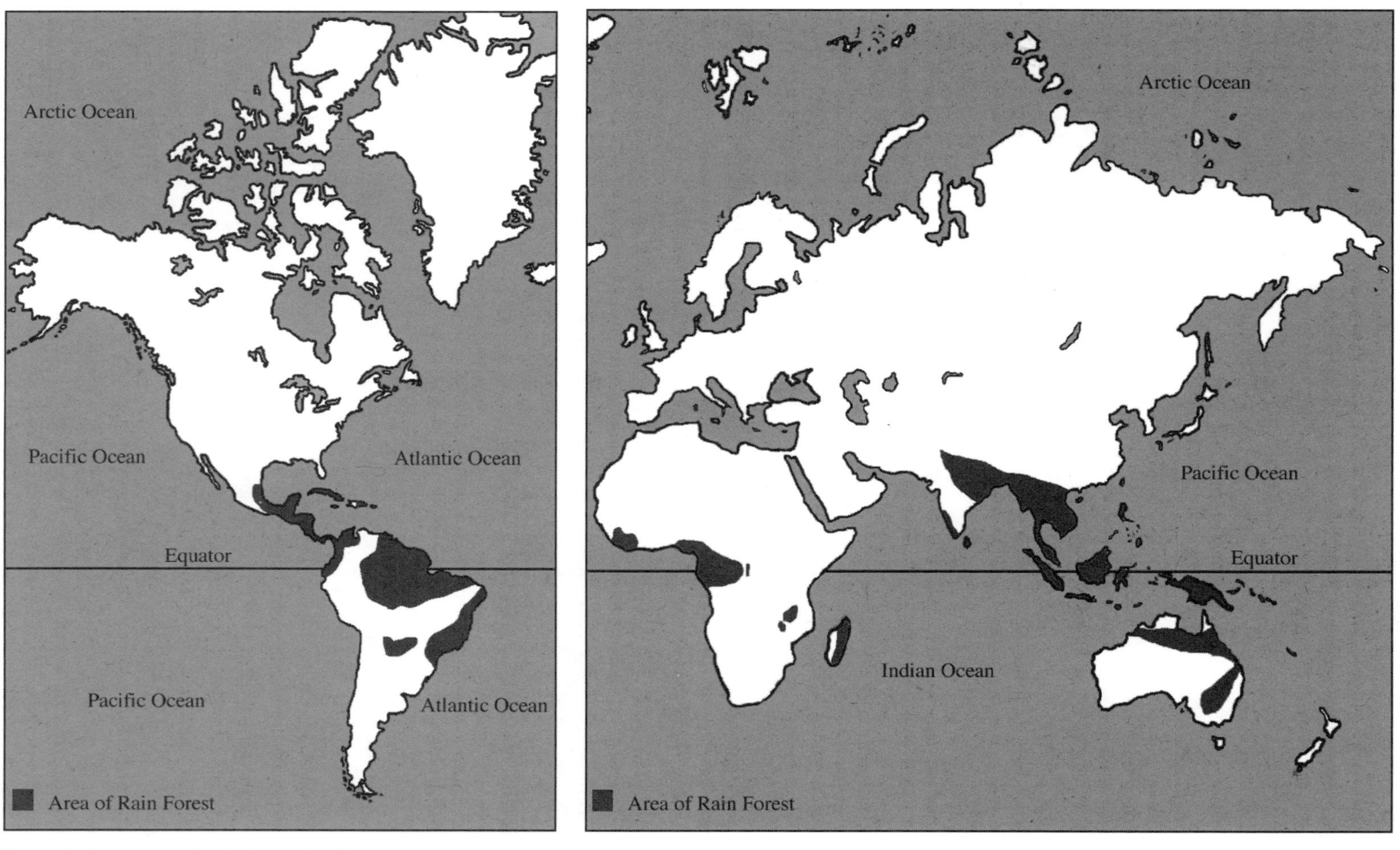

FIGURE 5-4 • **Areas of Rainforest in the Eastern and Western Hemispheres**

Some Rainforest Trees

Seventy percent of the 3,000 plants identified by the U.S. National Cancer Institute as having potential anticancer properties are found in the rainforest. One particular anticancer drug is Uña de Gato (Cat's Claw). The plant is a large wood vine that is found in the Amazon Rainforest and in other tropical areas. The vine can grow more than 33 meters (100 feet) around tree trunks. It is used in the treatment of dysentery, arthritis, and rheumatism. However, much attention is focusing on the chemical in the bark and roots of the plants. The chemical in Cat's Claw is used in the treatment for cancer and seems to be effective against the virus that causes *AIDS* as well as other diseases of the immunological system. It may also help prevent blood clots in blood vessels and lower blood cholesterol.

Another plant that shows promise as a treatment against cancer cells is Graviola. Graviola is a small evergreen tree that grows about 5 to 6 meters (15 to 20 feet) tall and is found in the tropical areas of South and North America. The natural chemicals in its leaf, bark, and twigs has been documented to possess antitumor properties.

Chemicals in other tropical plants are used in the treatment of Parkinson's disease, multiple sclerosis, and other muscular diseases. Quinine, an extract from the bark of the Cinchona tree, was an aid in the cure of malaria. The Cinchona comprises about 40 species of trees that grow up to 15 to 20 meters (45 to 60 feet). They grow in tropical regions, particularly on the eastern slopes of the Andes from Colombia to Bolivia. Today synthetic drugs have replaced the use of Quinine, but the bark extract is still used for food products and as a herbal medicine to treat leg cramps, as well as colds and flu, and for normalizing heart functions.

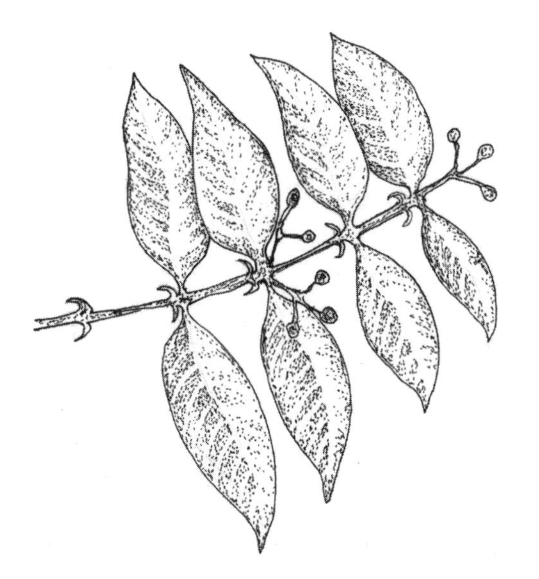

Figure 5-5 • The chemical in Cat's Claw, a large vine found in the rainforest, is used in the treatment of cancer and AIDS, as well as other diseases.

FIGURE 5-6 • Graviola is a small evergreen tree that grows in the tropical forest areas of North and South America. The natural chemicals in this plant show promise as an anti-tumor medicine.

FIGURE 5-7 • Quinine is a chemical that is an extract from the bark of a Cinchona tree that grows in the tropical forests of the Andes from Bolivia to Colombia. Quinine is used today as a herbal medicine to treat colds and flu.

The Brazilian Peppertree, found in Central and South America, is a small shrub that grows about 4 to 8 meters (12 to 24 feet) tall. The leaves, bark, and, fruit of the plant have been used to make commercial products such as syrups, resin, pepper, and beverages. As a medicinal plant, the peppertree is used to treat bronchitis, cataracts, gout, coughs, and fevers.

Suma is a large, shrubby ground vine found in the tropical rainforests of Brazil, Venezuela, and Peru. It is used in the treatment of asthma, leukemia, and high blood pressure. Chanca piedra is a small erect annual herb that grows about 30 to 40 centimeters (12 to 15 inches) and is found in the rainforests in the Amazon and in other tropical areas of

FIGURE 5-8 • Suma is another vine that grows in the tropical forests of Brazil, Venezuela, and Peru. The chemical in the plant is used to treat asthma, leukemia, and high blood pressure.

A Brazilian indigenous rubber tree worker explains how he extracts sap from a rubber plant. (Courtesy of Jane Mongillo)

the world. The chanca piedra extract is used in the treatment of kidney stones and gallstones. From Africa, Madagascar's pink-petaled rosy periwinkle plant provides two important antitumor chemicals that are used in the treatment of lymphocyctic leukemia in children and Hodgkin's disease patients. Before this drug was discovered, only 20 percent of the children suffering from leukemia recovered.

Today, *ethnobotanists* are researching medicines in rainforest plants. These scientists research and collect plant samples. Besides that work, they also talk with and record information from local *indigenous* tribes who have been using plant medicines for a long time. Several tribes in the Amazon have used the sap, bark, fruit, seeds, flowers, and roots of native plants to treat coughs, bronchitis, stomach problems, athlete's foot, diabetes, asthma, gallstones, hypertension, and many other illnesses.

Only about 1 percent of the rainforest plants have been tested and analyzed as possible medicines. What kinds of future drugs will be discovered from these plant species to treat AIDS, heart disease, and cancer? Most ethnobotanists agree that the future looks very promising. However, to obtain those dreams through plant research, it becomes very important to preserve the biodiversity in the tropical areas.

ENVIRONMENTAL CONCERNS OF FORESTS

Worldwide, forests continue to disappear at an alarming rate, with 50 percent of the world's original forest cover already lost to the pressures of logging, forest fires, and land clearing. Over 8 billion hectares (about 19 billion acres) of forest existed in the world 8,000 years ago; today, only 3–4 billion hectares (about 10 billion acres) remain. The pace of forest destruction accelerated in the 1990s and continues to rise; currently, over 400,000 hectares (950,000 acres) of forest are cleared or degraded every week. Given this alarming rate, increasing numbers of people are concerned about forest loss and forest management. Recent surveys show that the prospect of forest loss has overtaken recycling and chemical use as a key environmental issue for the

Indigenous People of the Rainforest

Many of the world's indigenous people live in the rainforests of South America. For the most part, they maintain their traditional life style and depend on the rainforest for food, shelter, and medicines.

There are about 50 million indigenous people who live in the tropical rainforests of the world. In South American alone there are many tropical forest dwellers. For example, about 70,000 indigenous people live on 6 million hectares (about 15 million acres) of land in the Colombian rainforest of South America. Colombian law gives these people the right to follow their own customs and tribal traditions.

Another indigenous tribe includes the Yanomami (Yanos), who live on 9 million hectares (about 23 million acres) of rainforest land in Brazil. The Yanos, approximately 11,000 people, live in doughnut-shaped houses. Each family has its own garden to grow vegetables and fruit. They also fish and hunt. The Kayapo live in the Xingu National Park in the Brazilian rainforests. They were among the first human beings to live in this part of Brazil. The 4,000 inhabitants live in 14 villages, where they raise crops such as sweet potatoes, fruit, and cotton in communal gardens.

general public. More than 8,750 of the 80,000 to 100,000 tree species known to science have been found to be threatened with extinction, and 77 are already extinct.

Refer to Volume IV, for more information about deforestation.

Vocabulary

AIDS (acquired immune deficiency syndrome) A viral disease that leads to the destruction of the human body's normal immune system and makes the body vulnerable to infections

Cellulose A carbohydrate that makes up a large bulk of a plant

Chemical An element or a compound of a definite composition that may exist as a solid, liquid, or gas

Deforestation Excessive cutting down and removal of trees

Developing nations Countries in which the gross domestic product is less than $7,000 per capita. These countries populate Latin America and Africa.

Ethnobotanist A scientist who studies the plant lore of native or indigenous people

Indigenous Living naturally in an area; native

Old growth forests A late stage in forest ecological succession in which there are many large, mature trees and often several canopy layers

Temperate A climate area that is neither too hot nor too cold

Watersheds A total land area that drains directly into a stream or river

Activities for Students

1. Why is it that developing nations use mostly wood as their fuel source? What are agencies and nongovernmental organizations like the UN, the Organization of Petroleum Exporting Countries, the Peace Corp, and the Kellogg Foundation doing to support or change this?
2. Which major universities have forestry schools geared specifically to preserve the woodlands of the world? Where are they located, and what courses do they offer?
3. How cost effective is paper recycling?
4. Pretend you are an ethnobotanist interested in studying the health effects of the flowers, fruits, and sap of the flora in Acre, Brazil. Write a grant for a 6-week study in the area and describe the challenges that would you face in your travels to the area and your cohabitation and communication with the Yanomani indigenous tribe in the area as you conduct your research.

Books and Other Reading Materials

Breymeyer, A. I., D. O. Hall, and J. M. Melillo, eds. *Global Change: Effects on Coniferous Forests and Grasslands*. Scope, No. 56. New York: John Wiley & Sons, 1997.

Jukofsky, Diane. *Encyclopedia of Rainforests*. Phoenix, Ariz.: Oryx Press, 2002.

Sensel, Joni. *Traditions through the Trees: Weyerhaeuser's First 100 Years*. Documentary Book Publishing Corporation, 2000.

Terborgh, John. *Diversity and the Tropical Rain Forest*. Scientific American Library, No. 38. New York: W. H. Freeman & Co., 1992.

Young, Allen M. *The Chocolate Tree: A Natural History of Cacao*. Smithsonian Nature Books. Washington, D.C.: Smithsonian Institution Press, 1994.

Websites

American Forests, http://www.treelink.org

American Forest & Paper Association (AF&PA), http://www.afandpa.org/

Forest Products Society (FPS), http://www.forestprod.org

Greenpeace International, Forests, http://www.greenpeace.org/~forests

National Association of State Foresters, http://www.sso.org/nasf/nasf.html

Paper Industry International Hall of Fame, http://www.paperhall.org/

The Rainforest Alliance, http://www.rainforest-alliance.org

Society of American Foresters, http://www.safnet.org

U.S. Forest Service, http://www.fs.fed.us

World Conservation Monitoring Centre, http://www.wcmc.org.uk

World Resources Institute Forest Frontiers Initiative, http://www.wri.org/ffi

World Wildlife Fund (Worldwide Fund for Nature) Forests for Life Campaign, http://www.panda.org/forests4life

Water: A Basic Resource

Renewable resources are those natural resources that are regularly replenished through natural processes and thus have the potential to last indefinitely if managed responsibly. Examples of such resources include saltwater and freshwater, which are naturally replenished through the water cycle.

Freshwater is one of the most precious basic natural and renewable resources. Human health and survival depends on a daily intake of clean water. The *cells* in the human body are full of water. Water has the ability to dissolve many substances in the body and thus allows the cells to use valuable nutrients, minerals, and chemicals to maintain body functions.

Oceans play a major role in the amount of precipitation that falls on Earth and have significant impacts on climate. But oceans also provide food and minerals.

DID YOU KNOW?

The human body, like the bodies of most animals, contains approximately 65 percent water by weight.

FRESHWATER USES

About 97 percent of Earth's water is found in the oceans. The water is salty and unusable for human consumption. The remaining 3 percent is freshwater, but most of it is locked up in the frozen ice of glaciers and the continental ice sheets of Greenland and Antarctica. The following analogy will illustrate how little of Earth's freshwater is available: If all of Earth's water could fill a 4-liter (about one gallon) container, the amount of available fresh water would fill less than a tablespoon out of the

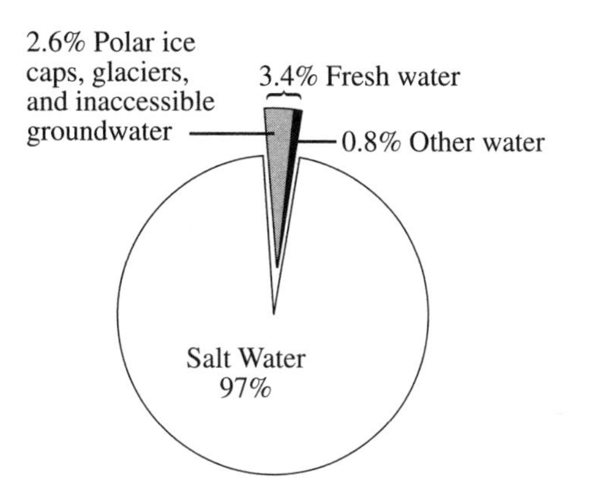

FIGURE 6-1 • The oceans, ice caps, and glaciers tie up more than 99 percent of total water in the environment. All of this water is unsuitable (ocean waters) or unavailable (ice caps, glaciers) for human use.

Water's Physical Properties

Water is unique in that it is the only natural substance that is found in all three states—liquid, solid (ice), and gas (steam)—at the temperatures normally found on Earth. Earth's water is constantly interacting, changing, and moving. Water freezes at 0° C (32° F) and boils at 100° C (212° F) at sea level. Water is unusual in that the solid form, ice, is less dense than the liquid form; this explains why ice floats.

Water can absorb a lot of heat before it begins to get hot. This makes water valuable to industries as a coolant. Water has a very high surface tension. In other words, water is sticky and elastic, and it tends to clump together in drops rather than spread out in a thin film. Surface tension is responsible for capillary action, which allows water, along with its dissolved substances, to move through the roots of plants and through the tiny blood vessels in our bodies.

Water is also called a universal solvent. That is, it dissolves more substances than any other liquid. This means that wherever water goes, either through the ground or through our bodies, it takes along valuable chemicals, minerals, and nutrients. Pure water has a neutral *pH* of about 7, which is neither acidic nor basic.

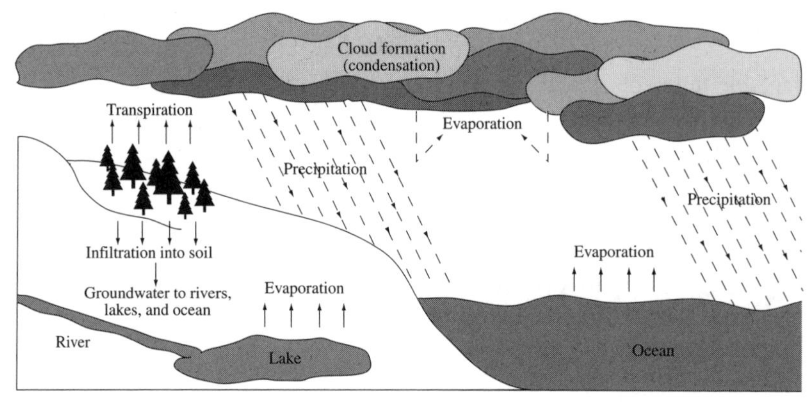

FIGURE 6-2 • Water Cycle The model of the hydrologic cycle shows the natural pathway through which water moves between Earth's surface and atmosphere and between living and nonliving things.

four liters. As you can see, only a small percentage of freshwater on Earth is available for human use.

The Water Cycle

The oceans are the ultimate source for all of our global freshwater supply. Earth's water moves constantly in a cycle between ocean, atmosphere, and land. The movement of the water in the cycle is known as the hydrologic cycle. Some water cycles are long and complex. Others are short and direct. Simply stated, the process includes ocean water evaporating in the heat of the sun. The water vapor rises into the atmosphere, forming clouds that deposit precipitation as they move over the continents and oceans. Much of the precipitation falls directly into the ocean. On land, some of the precipitation that reaches Earth's surface flows as surface water into streams and rivers that flow downward into lakes or toward the ocean. But not all the precipitation that falls on Earth runs off into streams or rivers. Some of the water seeps into the

TABLE 6-1 Selected Water Quality Standards

Substance	Limit
Arsenic	0.05 parts per million (ppm)
Carbon tetrachloride	0.005 ppm
Copper	1.3 ppm
Cyanide	0.2 ppm
Lead	0.015 ppm
Nickel	0.14 ppm
Coliform count	Less than 10 colonies per liter
pH	6.5–8.5

Source: U.S. Environmental Protection Agency, National Primary and Secondary Drinking-Water Standards.

ground. Water that occurs beneath Earth's surface at depths of a few centimeters to more than 300 meters (900 feet) is called groundwater.

Freshwater Sources

Surface water and groundwater are the two major sources of freshwater supplies. Freshwater that is safe for use by people as drinking water is called potable water. Potable water must be fresh—that is, have a salinity lower than 0.35 grams per liter—and be free of contamination by pollutants or pathogens. Potable water is also generally free of minerals that give the water a bad taste, make it too acidic or too alkaline, or produce an objectionable odor. In the United States, the Environmental Protection Agency (EPA) has set limits, called maximum containment levels, for the amounts of various substances in potable water.

Treating Drinking Water for Safety

Many cities get their fresh water supplies from surface water in rivers, lakes, dams, and reservoirs. Before the water is available to the public, it needs to be treated at a treatment plant to ensure that the water is of high quality and safe. Several steps are used to treat the water. First, screens are used to keep out leaves, fish, and other large objects. After this process, the water is treated with alum and other chemicals. The alum in the water forms a sticky substance that removes impurities, such as algae, in clumps. Eventually the clumps become heavy and sink to the bottom of settling tanks, where they are removed. The water then flows into a sand-and-gravel filter that removes more impurities. The next step is to add chlorine, which is used to kill any disease-causing organisms. After the chlorination process, the water is aerated to remove odors and unpleasant tastes. The last step in the treatment process may include adding *fluoride* to the water. The water is then distributed to communities and businesses.

DID YOU KNOW?

The Romans built aqueducts more than 2,000 years ago to carry water from its source to different locations. The remains of some of these aqueducts can be seen today in France, Spain, and other countries.

SAFE DRINKING WATER ACT

The Safe Drinking Water Act became federal law in 1974. The law was passed in response to outbreaks of waterborne disease and increasing chemical contamination of public drinking water in the United States. This law focuses on protection of all waters that are actually or potentially suitable for public drinking use. This includes water from both above-ground and underground sources. The act protects drinking water supplies by establishing water quality standards for drinking water, monitoring public water systems, and guarding against groundwater contamination. Before 1974, each state set up its own drinking water program; as a result, drinking water protection standards differed from state to state.

Refer to Volume IV for more information about water pollution.

ENVIRONMENTAL CONCERNS OF WATER The demand for potable water often exceeds the supply in remote areas and developing nations. In such places, people usually rely on surface water for their water supplies. This water may be shared with wildlife, which use the water as habitat, for drinking, or for cooling, and it may be contaminated with pathogens or be unclean because of suspended particulates or algae. Despite these problems, people may be forced to use such water for drinking, cooking, and bathing because it is the only available water. Use of unclean water is responsible for the transmission of many diseases. For example, giardiasis is an intestinal disease caused by ingesting water containing the organism *Giardia lamblia,* a protozoan. Other serious diseases such as cholera and typhoid are also transmitted in contaminated water. Infection from such pathogens can often be avoided by boiling water to kill them.

TABLE 6-2 Water Used in the Home

Task	Water Used (liters)
Showering for five minutes	95.0
Brushing teeth	10.0
Washing hands	7.5
Flushing standard toilet	23.0
Flushing "low-flow" toilet	6.0
Washing one load of laundry	151.0
Running dishwasher	19.0
Washing dishes by hand	114.0

TABLE 6-3 Uses of Surface Water and Groundwater (percentage)

Irrigation	39
Thermoelectric power	39
Public Supply	12
Industry	6
Livestock	1
Domestic (Home use)	1
Mining	1
Commercial	1

GROUNDWATER

Public water supplies from groundwater usually require less steps in the treatment process than water originating from surface waters. The reason is that much of the underground water flows through sand-and-gravel beds that act as natural filters. These filters help remove bacteria and other microorganisms. However, groundwater for domestic use is sometimes treated with chlorine.

Groundwater moves through layers of porous materials called aquifers. An aquifer is a natural underground water resource that is an important source for drinking water and irrigation in many parts of the world. Aquifers may consist of soil, which is made up of separate grains, or rock containing fractures or other channels. In an aquifer, the groundwater flows through the spaces between the soil particles or within the rock openings. Within many aquifers, groundwater flows horizontally through the natural spaces in the soil or rock until it reaches a discharge area, such as an ocean, lake, or stream, where the water reaches the surface of the land.

Aquifers also include recharge areas where water enters the ground and travels downward into the aquifer, replenishing the supply of water. Recharge areas may include surface water bodies or wetlands. Aquifers store large quantities of water below ground, where the water is better protected from effects of *contaminants* that exist at the surface. However, aquifers require protection and careful planning to ensure that they do not become polluted by groundwater contamination as a result of inappropriate land uses or accidental chemical leaks or spills.

Aquifers are one of Earth's most important natural resources. Groundwater is the source of about 38 percent of the water that county and city water departments supply to households and businesses (public supply). It provides drinking water for more than 97 percent of the rural population who do not get their water delivered to them from a county or city water department or a private water company. Even some major cities, such as San Antonio, Texas, rely solely on groundwater for all their needs. About 37 percent of all water sources used for irrigation comes from groundwater.

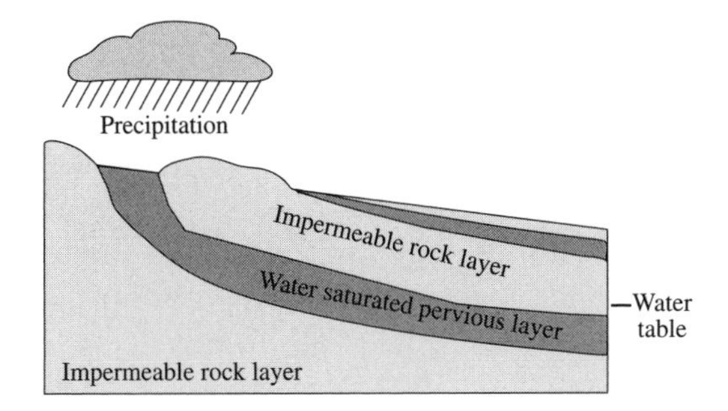

FIGURE 6-3 • Aquifer
An aquifer is a natural underground water resource that is an important source for drinking water and irrigation in many parts of the world.

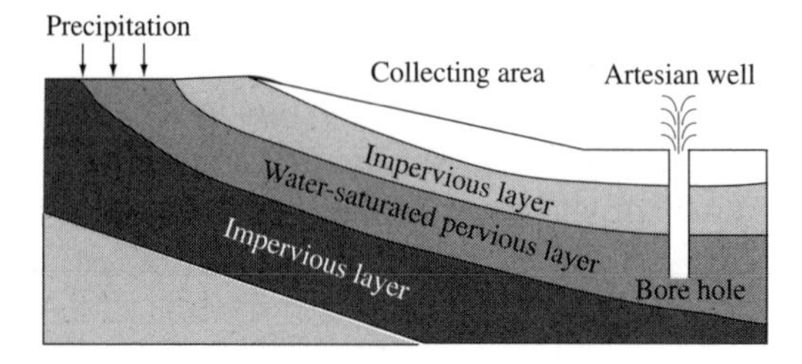

FIGURE 6-4 • Artesian Well The artesian well allows groundwater to rise to the surface without mechanical pumping.

WELLS Generally water is obtained from aquifers through wells that are drilled or dug into the ground. If the well allows groundwater to flow naturally to the surface without mechanical pumping, it is known as an artesian well. However, most groundwater is usually pumped out of an aquifer from a well. The process of removing groundwater from an aquifer is called withdrawal, and the rate of withdrawal is often measured in cubic meters per day.

The two principal types of water wells are dug wells and drilled wells. Dug wells consist of a hole dug in the ground and lined with rocks or other materials that keep the walls of the hole from collapsing. Dug wells are very close to the surface. The drilled wells are constructed with special drilling equipment and may go down several hundred meters or more. Water is removed from both types of well by pumps.

ENVIRONMENTAL CONCERNS OF GROUNDWATER Not all groundwater is suitable for drinking water or industrial use. Groundwater from deep within Earth, such as groundwater pumped in oil fields (generally relatively old groundwater) may be highly saline (salty). In urban or industrial areas, often groundwater has become degraded from the results of human activities. The resulting groundwater pollution may make the water unsuitable for drinking without treatment. In the United States, state environmental regulations usually define standards of groundwater quality; these standards are based on the chemical properties of the groundwater that indicate whether it may be contaminated or otherwise unfit for consumption. Often states also delineate and map areas of groundwater protection (aquifer protection) and known or suspected groundwater degradation.

Water usage from aquifers must be closely monitored to avoid overdraft—a condition that results when water is removed from a source more quickly than it can naturally be replenished. When overdraft occurs, lands above an aquifer may subside and compact (sink) to fill the space once occupied by the water. Aquifers located near coastlines also risk contamination with salt water from the sea, a condition known as saltwater intrusion, which withdrawals or overdraft may exacerbate.

Use of Water for Irrigation

Irrigation is the process of artificially supplying water to agricultural land used for growing crops. The history of irrigation goes back to approximately 4000 B.C. In countries such as India, Egypt, and China, for example, irrigation has been practiced for thousands of years. Irrigation was also practiced in South America. About 1,500 years ago, people built hundreds of kilometers of irrigation canals in the Andes mountains of Peru. The water was used to grow corn, quinoa, potatoes, grains, and alfalfa.

Both groundwater and surface water are used for irrigation. According to the United Nations Food and Agricultural Organization (FAO), about 17 percent of the world's cropland used some form of irrigation in 1994. This land, which represents almost 250 million hectares (618 million acres) of cropland, produces almost 40 percent of the world's food.

Methods of Irrigation

Worldwide, there are several types of irrigation in use. They include flood irrigation, furrow irrigation, spray or sprinkler systems, and drip (or trickle) irrigation. The method used on a particular farm is determined by such factors as crop type, topography, water source, soil drainage, and cost.

In 1995, about 32.7 million acres of the 59.3 million total irrigated acres (about 55 percent) were irrigated by the flood irrigation process. Another 24.9 million acres were spray irrigated, with the remaining 1.7 million acres received drip irrigation.

Flood Irrigation In flood irrigation, water is poured onto the land and flows through the field. Flood irrigation is most often used where land is relatively flat, having less than a 3 percent slope. Flood irrigation is also used where the water source is nearby and the crops being

TABLE 6-4 Irrigated Area in the United States, Top 20 States, 1997

1.	California	11.	Washington
2.	Nebraska	12.	Wyoming
3.	Texas	13.	Utah
4.	Arkansas	14.	Mississippi
5.	Idaho	15.	Arizona
6.	Colorado	16.	Louisiana
7.	Kansas	17.	Missouri
8.	Montana	18.	New Mexico
9.	Oregon	19.	Nevada
10.	Florida	20.	Georgia

grown require large amounts of water. Flood irrigation is among the least expensive irrigation methods because it does not require much machinery to carry and deliver water to plants. However, flood irrigation should be used only in areas having good soil drainage to avoid water-logging and salinization, which occurs as water evaporates from the land and leaves behind salts and minerals that accumulate in the soil. A drawback of flood irrigation is that this method wastes great amounts of water because much of the water goes to land not occupied by plants. In addition, more than 50 percent of water may be lost to evaporation.

FURROW IRRIGATION The delivery of water to crops via small channels, or furrows, dug between crop rows is called furrow irrigation. Furrow irrigation is used in areas with fairly flat topography. Because this method delivers water nearer to plants, furrow irrigation is more efficient than flood irrigation; however, furrow irrigation is also more expensive than flood irrigation. To offset the higher costs, this irrigation method is used for crops with high market values, such as cotton and vegetables. The main disadvantages of furrow irrigation is that it requires great quantities of water, much of which is lost to evaporation. The high evaporation rate may decrease soil fertility through salinization.

SPRAY OR SPRINKLER SYSTEMS Spray irrigation is water that is sprayed from spray guns onto fields. The use of sprinkler systems to deliver water to crops is a type of overhead irrigation useful in regions with all kinds of topography. Sprinkler systems are more expensive than flood or furrow irrigation because they require the purchase of equipment that delivers water from the air, simulating rainfall; however, the system allows water to be directed where it is needed and wastes less water than flood or furrow irrigation. Thus, sprinkler systems are useful in regions with limited water supplies. A drawback to sprinklers is that their efficiency is decreased by strong winds that divert water away from crops.

DRIP IRRIGATION The technology for drip or trickle irrigation was first developed in Israel and is today widely used throughout the United States, Israel, and Australia. Drip irrigation delivers water through narrow tubes directly to the root area of individual plants at frequent intervals and in small amounts. The slow, frequent release of water has several advantages over other irrigation methods. First, it reduces the total amount of water needed to irrigate crops, making this irrigation method useful in regions with limited water supplies. Second, because water is released slowly and over a small area, little water is lost to the air through evaporation or to the soil via percolation. At the same time, fewer salts are deposited in the soil, helping to preserve soil quality. Third, drip irrigation can be used on almost all lands, regardless of their topography. In fact,

Drip irrigation allows a slow and frequent release of water over the soil which helps prevent runoff and slow down the rate of evaporation loss. (Courtesy of USDA, NRSC)

TABLE 6-5 Irrigated Area, Top 20 Countries, 1994

1.	India	11.	Uzbekistan
2.	China	12.	Spain
3.	United States	13.	Iraq
4.	Pakistan	14.	Bangladesh
5.	Iran	15.	Egypt
6.	Mexico	16.	Romania
7.	Russia	17.	Brazil
8.	Thailand	18.	Afghanistan
9.	Indonesia	19.	Japan
10.	Turkey	20.	Italy

drip irrigation works well even on steep slopes. The main disadvantage of drip irrigation is that the cost involved in setting up the system is higher than that of other irrigation methods.

ENVIRONMENTAL CONCERNS WITH IRRIGATION

Irrigation greatly increases the amount of food that can be grown in an area. However, use of irrigation is declining in many regions because of problems such as *salinization* of soil and water shortages. Salinization occurs slowly as water evaporating from soil leaves behind mineral deposits that build up and render the soil infertile.

Pivot irrigation is used on a farm in southern Utah. (Courtesy of Hollis Burkhart)

Ogallala Aquifer

The largest aquifer and groundwater source in North America, the Ogallala Aquifer is located in the midwestern United States, beneath portions of South Dakota, Nebraska, Colorado, Kansas, Oklahoma, New Mexico, Wyoming, and northern Texas. The underlying aquifer is approximately 440,000 square kilometers (170,000 square miles). Most of the volume of water is under Nebraska. Also known as the High Plains Aquifer, this great groundwater resource provides irrigation water for approximately 10 million acres (4,005,000 hectares) of farmland.

The Ogallala Aquifer has an average thickness of about 61 meters (200 feet), consisting largely of gravel, sand, and silt deposits carried eastward from the Rocky Mountains. These water-bearing deposits were later buried by other geologic processes.

The quantity of water stored within the Ogallala Aquifer is estimated to equal the volume of Lake Huron, one of the Great Lakes. The groundwater level beneath the Ogallala Aquifer has been dropping since the 1940s, which signifies that the amount of water withdrawn (pumped) from the aquifer is greater than the amount of water entering (or recharging into) the aquifer. Some environmentalists believe that the aquifer will be completely depleted by 2020 as a result of the amount of water being withdrawn. As an example, in parts of Texas and New Mexico, water levels in the aquifer have declined by more than 30 meters (90 feet) since water-pumping operations started in the 1940s. Water levels in the aquifer under Kansas and Oklahoma have also declined. This indicates the aquifer is shrinking and not being renewed. As a result of the depletion of water in the aquifer, farming areas in parts of Texas and New Mexico have cut back on using the aquifer for irrigation purposes.

OCEANS: SALTWATER RESOURCES

More than 70 percent of Earth is covered by oceans. The ocean is a valuable source of food, chemicals, and minerals.

FIGURE 6-5 • Ogallala Aquifer The Ogallala aquifer in mid-Western United States is the largest aquifer and groundwater source in North America.

Food Sources from the Ocean

People have been taking fish from the ocean for thousands of years. Laws relating to fishing methods were written into English law as early as A.D. 400. Today commercial fishing is an industry in which fish are harvested, either in whole or in part, for sale or trade. Finfish and shellfish are among the economically dominant features of the world's oceans, as well as vital sources of protein for the world's people. Finfish are those animals that have a spine, bones, and fins for swimming. The major finfish include cod, haddock, flounder, salmon, tuna, herring, and halibut. Shellfish are invertebrates—those animals without backbones. They have shells or shell-like material enclosing their bodies. Major shellfish include clams, oysters, squid, shrimp, lobster, mussels, and octopus.

Since ancient times, fishing has been a provider of employment and economic benefits to those engaged in this activity. Worldwide, commercial landings of fish nearly quintupled between 1950 and 1989, from 20 million metric tons to nearly 100 million. But by the year 2010, world demand for edible seafood is projected to be 110 million to 120 million metric tons.

ENVIRONMENTAL CONCERNS OF OCEAN FISHING

For most of our history, the wealth of aquatic resources has been assumed to be unlimited. By the late 1960s, however, it became clear that fishery resources were beginning to decline. Since then, in some regions of the world declines of commercially important fish stocks have

TABLE 6-6 Commercial Fishing

Region	Gross Tonnage (1,000 metric tons) 1970	1992	Percent Growth 1970–92
Asia	4,802.3	11,012.5	129
Former Soviet Union	3,996.7	7,765.5	94
Europe	3,097.4	3,018.3	−3
North America	1,076.9	2,560.0	138
South America	361.5	816.5	126
Africa	244.0	699.1	187
Oceania	37.1	122.3	230
World	**13,615.9**	**25,994.2**	**91**

become so severe that the welfare of coastal communities and regions has become threatened. Perhaps the most dramatic depletions of fish stocks have been in the western Atlantic Ocean, where commercially viable quantities of cod have all but vanished from the Grand Banks. Other environmental concerns include the destruction of coastal spawning habitats and fishing practices that kill immature fish and nontarget species.

To learn more about the decline of commercial fishing, refer to Volumes IV and V.

Aquaculture

The *cultivation* of fish, shellfish, or aquatic plants in natural or controlled marine or freshwater environments is known as aquaculture. Estimates report that currently about 20 percent of all commercial fish are raised in an aquaculture environment, and that industry will continue to grow in the twenty-first century. Today aquaculture is a multi-million-dollar business. Much of the trout, catfish, and shellfish consumed in the United States are products of aquaculture.

Practiced since ancient times, aquaculture today encompasses a wide variety of activities, including the cultivation of fish, such as catfish and trout for food; the rearing of ornamental fish, such as carp and koi for aquariums; the raising of bait fish for the fishing industry and sporting fish for restocking lakes and ponds; the cultivation of oysters for obtaining pearls; the cultivation of mussels for food; and the growing of seaweed for food. The science of aquaculture is far behind that of its terrestrial counterpart—agriculture. Worldwide, however, aquaculture has grown dramatically in the past 20 years, with many species of fish and shellfish being produced.

Fish Aquaculture

The practice of cultivating and raising fish for food probably began as early as 4,000 years ago in China. Today fish farming, an important industry in the United States, the Philippines, Japan, China, India, Israel, and Europe, is by far the most common form of aquaculture. Fish farming is the practice of raising fish in captivity to improve their growth and reproduction, similar to the way livestock are raised on land. Most fish farms consist of many enclosures, ponds, lakes, tanks, pens, and long, narrow channels, each containing fish at varying stages of development. Fish culturists manage the aquatic environments by circulating clean water through the enclosures and protecting the fish from predators, disease, and *parasites*. In these types of farm operations, fish are grown to maturity and then harvested for food.

Another type of fish farm common in the United States is known as a fish ranch. In a fish ranch, many fish species, particularly sport fish species, such as salmon, are hatched in small ponds and then released

Tuna

Tuna is the common name for several large marine game and food fishes of the mackerel family. Tuna live throughout much of the world's oceans and are economically important to many countries. The United States, Nova Scotia, Japan, Thailand, the Philippines, Brazil, Colombia, Australia, Portugal, and several other Mediterranean countries have large tuna industries. The most commercially important tuna belong to the genus *Thunnus,* which includes the albacore (a white-meat tuna), the yellowfin, and the bluefin. The skipjack (or striped) tuna, which belongs to the genus *Katsuwonus*, also has commercial importance.

Many countries now have quotas limiting annual tuna captures; however, populations continue to decline. Some scientists estimate that the Atlantic tuna population alone has decreased by as much as 90 percent over the last 30 years. If this trend continues, some tuna species may soon be in danger of extinction.

One problem resulting from tuna fishing relates to fishing methods. Traditionally tuna, which travel in large schools, have been captured using driftnets. This fishing method results not only in the capture of great numbers of tuna, but also in the capture and deaths of other marine organisms, including dolphins, porpoises, and sea turtles. In the 1970s, concerns over the capture of dolphins in tuna fishers' nets brought about public protests and boycotting of tuna products. In response to the boycotts, many tuna fishers now use nets that capture tuna while allowing dolphins and other nontuna species to escape. Companies that package tuna captured using the new nets often market their products in cans carrying "dolphin-friendly" labeling.

Another problem affecting tuna populations is pollution of ocean waters with chemicals. Some of these chemicals are directly toxic to tuna and other marine organisms; others, such as mercury, collect in the tissues of organisms and increase in concentration as they move through the food chain in a process called bioaccumulation.

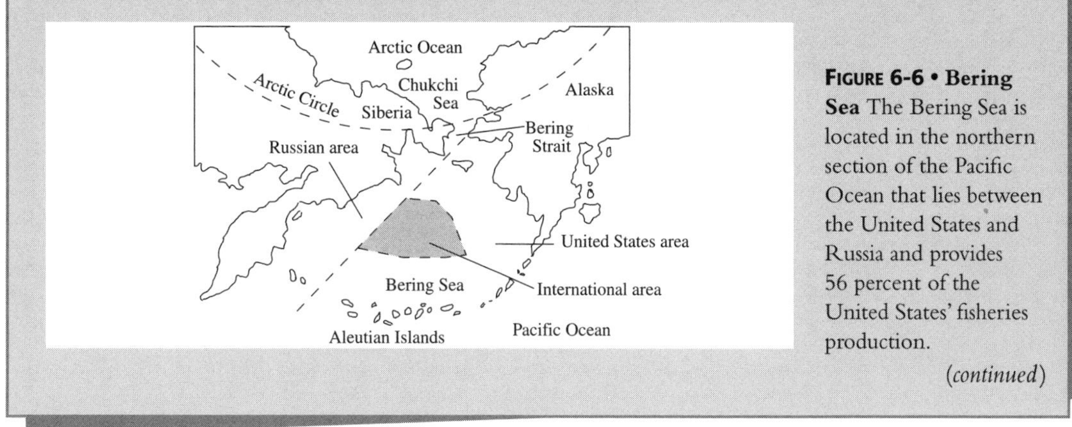

FIGURE 6-6 • Bering Sea The Bering Sea is located in the northern section of the Pacific Ocean that lies between the United States and Russia and provides 56 percent of the United States' fisheries production.

(continued)

Tuna (*continued*)

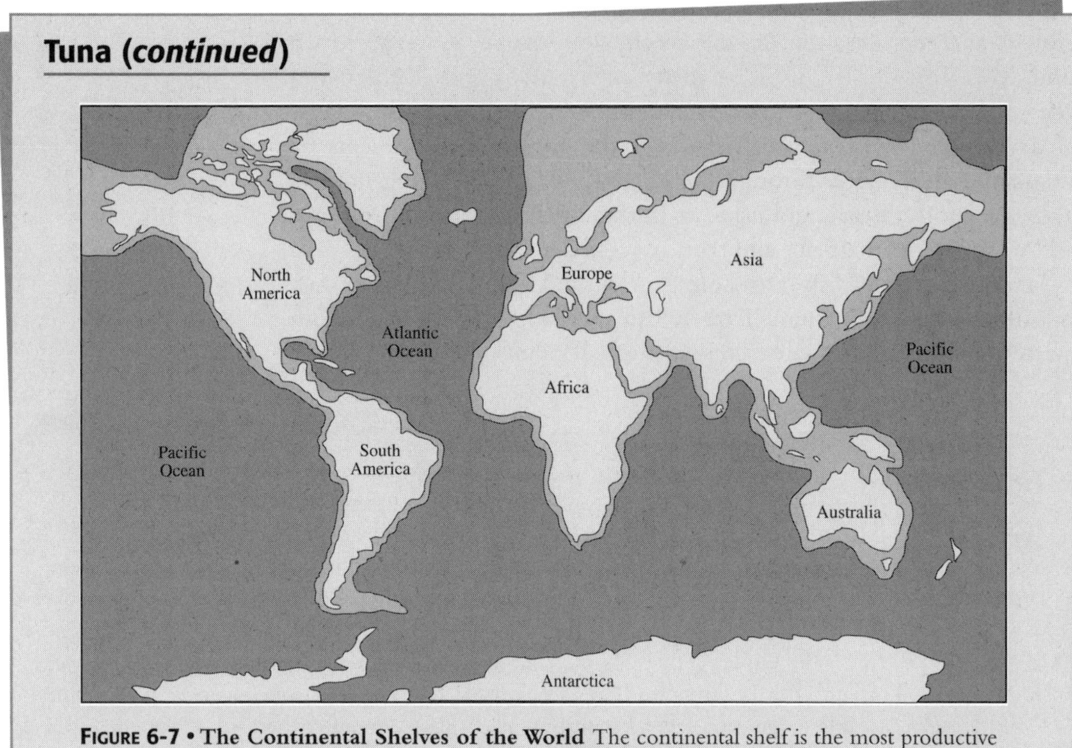

FIGURE 6-7 • The Continental Shelves of the World The continental shelf is the most productive area of the ocean with respect to commercial fishing and offshore oil drilling.

TABLE 6-7 World Aquaculture

Country	Percent Share of Global Production
China	57
India	9
Japan	4
Indonesia	4
Thailand	3
United States	2
Philippines	2
Korea, Republic of	2
Other Countries	17

into rivers. The fish then migrate downstream to the ocean, where they will reach adulthood. Once these fishes mature, they instinctively return to the river from which they were released in order to reproduce. When they do so, they are captured and harvested for food.

PLANT AQUACULTURE

The aquatic plants raised in aquaculture include ornamental plants, such as pond lilies, and native species of plants used for habitat restoration. However, the vast majority of the plants systematically grown in aquaculture operations are seaweeds, a type of algae. The cultivation of seaweed is particularly popular in China and Japan, where it is an important food source. Since the seventeenth century, for example, Japanese aquaculturalists have grown their own seaweeds. Traditionally farmers cultivated the algae by placing long bamboo sticks into rivers. When small seaweed plants began to grow, the sticks were removed and brought to the sea, where the plants thrived in a mixture of fresh and salt water. Today the cultivation of algae in Japan and elsewhere is highly mechanized. Small seaweed plants are first grown in special hatching tanks. The plants are then trucked to the coasts, where they are mechanically tended to until harvest.

There are several economic and social forces behind the tremendous growth and interest in aquaculture. The main reason, perhaps, is the recognition that the world's oceans, lakes, and rivers cannot produce enough food to satisfy the world's appetite for fish and other types of seafood. Therefore, any shortages in seafood can be made up only through aquaculture.

Another force behind the growth in aquaculture is the increased interest, particularly within the United States, about eating a healthful diet. Numerous studies have recognized that fish and seafood are low in sodium, fat, and *cholesterol*. Additional studies have found that certain fish contain fatty substances that have the effect of reducing cholesterol in the body.

DID YOU KNOW?

Idaho uses about 5,600,000 liters (about 1,400 million gallons) of water per day to grow trout and accounts for about 80 percent of the world's farm-raised trout. In Louisiana, more than 50 times more water is used for fish farming than is used for animals that produce meat, poultry, and milk.

ENVIRONMENTAL CONCERN OF AQUACULTURE

However, there are problems with some of the fish farms, particularly shrimp-raising farms. Disease outbreaks, chemical pollution, and the environmental destruction of marshes and mangroves have been linked with fish farm activities. Environmentalists believe more *sustainable* practices are needed to control the potential for pollution and damage of natural resources. There is also concern about the accidental releasing of cultivated fish into natural populations and what effects that might have.

Ocean Minerals

The ocean is most important for supplying marine food for human beings. In addition to its value as a source of food, the ocean storehouse also contains chemical elements. The most common is sodium chloride, or table salt. Salt has been taken from the ocean for centuries. Magnesium, manganese, and bromine are elements that are being extracted in increasing amounts from ocean water.

Magnesium (Mg) is a light metal that is combined with other metals to make alloys that are light but very strong. These alloys are used in the manufacturing of airplane parts, bicycles, and automobiles. In industry, manganese (Mn) is combined with iron to make steel and other alloys. Manganese compounds are also used to make a variety of products including *pesticides*, fertilizers, ceramics, paints, dry cells, disinfectants, antiseptics, and a dietary supplement that helps the body use vitamin B_1. Bromine is used in an additive in gasoline to help it burn better in engines. A common resource on the ocean bed are manganese nodules which have a high content of copper, nickel, and cobalt. Manganese is used in the production of alloys in making steel products.

Manganese nodules were first discovered on the ocean floor in 1803. Since the 1960s manganese nodules have been recognized as a potential ore source because of the depletion of land-based mineral resources. Ocean research surveys have found large deposits in the Okinawa Trough (between Taiwan and Kyushu) at depths of 1,600 meters (about 4,800 feet). Nodule fields seem to be widespread in the Pacific Ocean. The Japanese are developing an ocean miner that could be used for deep-ocean mining of manganese nodules as well as other materials.

Refer to Chapter 5 for more information on minerals and metallic minerals.

DESALINATION

The ocean can also supply freshwater in areas where this resource is in short supply. The process of removing dissolved salts from seawater and brackish water is called desalination, also known as desalinization. Desalination is a very important process in many drought-prone areas of the world because it can transform unusable water from the ocean into valuable freshwater.

The average salt content of ocean water is about 3.5 percent salt and that of brackish water is about 0.5 percent to 1 percent. Desalinization reduces the salt content of these waters to about 0.05 percent. There are two general methods for removing these salts: distillation and *reverse osmosis*. Distillation is the simpler and more inexpensive

A model of the Taunton River desalination plant to be built on the river in Massachusetts. The reverse osmosis treatment plant will provide 20 million liters of water a day using the brackish water of the lower Taunton River in southeastern Massachusetts. (Courtesy of Bluestone Energy Services, Inc.)

method. In distillation, freshwater is simply evaporated from saltwater, leaving the salts behind. In reverse osmosis, seawater is forced under high pressure through a filter, which traps the salt crystals. This method purifies about 45 percent of the saltwater that passes through the filter. The remaining, more concentrated saltwater is pumped back into the sea. Reverse osmosis is commonly used in very dry areas, such as the Persian Gulf, where freshwater supplies are limited.

Desalination became important during the nineteenth century when steam-powered ships were in use. People were searching for dependable supplies of freshwater that would not damage the steam engines. In 1869, the first patent for a desalination process was granted. That same year, the first desalination plant was built near the Red Sea by the British government to supply freshwater to ships that came to port. Today there are approximately 7,500 desalination plants worldwide. Most of these are located in countries around the Persian Gulf, such as Saudi Arabia and Kuwait, but a few smaller plants are now operating in southern California. Generally desalination is not used to produce a region's total water supply; instead, the process aims to supplement existing freshwater supplies or in emergency situations, such as after long droughts.

Although there are many desalination plants scattered throughout the world, together they produce less than 1 percent of the world's freshwater supply. Desalination is a very expensive process that requires a lot of energy. For example, in the United States, desalinized water produced by the reverse osmosis method can cost up to $3 per 3,800 liters (about 1,000 gallons). This is about four to five times what the average U.S. citizen pays for fresh drinking water, and it is over 100 times the price paid by farmers for irrigation water.

Because of the high costs, desalinized water is mainly used to supply drinking water in the home. Small reverse osmosis desalination units with a few liters-per-day capability can be purchased for home use. In areas such as the Persian Gulf, where energy is inexpensive, using desalinized water for these purposes is quite practical.

Vocabulary

Cells The structural and functional units of all living things, the presence of which is an indicator that something is an organism. Cells, such as those of multicellular plants and animals, may be specialized to carry out specific functions for the entire organism.

Cholesterol Fatty substances found in fats and oils; produced by the liver; high cholesterol levels in the body are associated with certain heart diseases.

Contaminants Substances that are harmful to organisms.

Cultivation The process of growing crops or other organisms.

Fluoride A chemical element that is added to the public water supply to prevent tooth decay.

Parasite An organism that lives on or in another organism and gains a benefit from the relationship.

Pesticides Chemicals that are used to kill unwanted organisms.

pH The measure of the concentration of hydrogen ions in a solution to determine if it is acid or alkaline. The pH factor is shown as a number. A value of 7 is neutral. Lower numbers indicate increasing acidity, whereas higher numbers show increasing alkalinity. The pH of 0 is most acid. The pH of 14 is most alkaline.

Reverse osmosis A process in which saltwater is forced through a membrane to remove salt and other impurities to provide less salty water.

Salinization A process by which the soil becomes more salty, usually where there is a high usage of irrigation.

Sustainable Does not deplete or damage natural resources; to keep in existence.

Activities for Students

1. Experiment with the freezing, condensation, and evaporation points of freshwater and saltwater as it might occur up and down the latitudes of Earth. How does the salt, soil, mineral, and plant content of the freshwater and saltwater induce or inhibit these processes?
2. Contact your local water agency for a water-testing kit. Test the water you drink for different compounds and minerals in the various places you live—at home, in school, at work, and in public places. Why is fluoride placed in some public water supply systems? Who makes that decision?
3. People have been digging for water since the beginning of time. Describe the various tools that the human race has developed in order to know where exactly to dig for water.
4. Dehydrating ocean water to produce salt played a major role in the liberation of India under Gandhi's leadership. Why and how?

Books and Other Reading Materials

Fields, Leslie Leyland. *Out on the Deep Blue: Women, Men, and the Oceans They Fish*. New York: St. Martin's Press, 2001.

Gleick, Peter H. *The World's Water, 2000–2001: The Biennial Report on Freshwater Resources (World's Water, 2000–2001*. Washington, D.C.: Island Press, 2000.

Lewis, Scott Alan. *The Sierra Club Guide to Safe Drinking Water*. San Francisco: Sierra Club Books, 1996.

Pielou, E. C. *Fresh Water*. Chicago: University of Chicago Press, 1998.

Postel, Sandra, *Last Oasis: Facing Water Scarcity. Worldwatch Environmental Alert Series*. New York: W. W. Norton & Company, 1997.

Websites

EPA Website, http://www.epa.gov/swerosps/ej/, a site for innovative technologies for contaminated soil and groundwater.

National Oceanographic and Atmospheric Administration Fisheries, http://www.nmfs.gov/

NOAA Fisheries Contact website, http://www.nmfs.gov/

U.S. Department of Agriculture, website: http://www.usda.gov

U.S. Fish and Wildlife Service, U.S. Department of the Interior, http://www.fws.gov

U.S. Geological Survey, http://www.water.usgs.gov/

United Nations Food and Agriculture Organization Fisheries, http://www.fao.org/waicent/faoinfo/fishery/fishery.htm

Wildlife and Wilderness Resources

Today many global government agencies and nongovernment organizations (NGOs) have set aside areas for the conservation, protection, and management of *wildlife*. These wildlife resource areas include wildlife refuges, national parks, and national forests, as well as national monuments and historic areas. Other wilderness sites include wild rivers and recreational trails.

WILDLIFE REFUGES

Wildlife refuges are areas of land and water that provide food, water, shelter, and space for wildlife. Wildlife refuges are maintained usually by a government or nonprofit organization such as the Nature Conservancy, the Wilderness Society, and the National Audubon Society. Besides the United States, other countries throughout the world also maintain wildlife refuges.

National Wildlife Refuge System

The National Wildlife Refuge System is the name of the government wildlife refuge in the United States and is administered by the U.S. Fish and Wildlife Service. It is the world's largest and most diverse collection of lands specifically set aside for wildlife. Each refuge is a *sanctuary* for plants, animals, and other wildlife. By far the most numerous are the waterfowl refuges, which supply breeding areas, wintering areas, and resting and feeding areas along major flyways during migration. The major objective of the National Wildlife Refuge System is to conserve and protect the breeding and wintering grounds for migratory waterfowl such as ducks and geese and to preserve habitats for endangered species. In many places, including the prairie pothole region, each year thousands of hectares (acres) are planted with native grasses improving waterfowl habitat.

The National Wildlife Refuge System was begun in 1903 by President Theodore Roosevelt. He designated Florida's 2-hectare (5-acre) Pelican Island to protect and save egrets, pelicans, herons, and other birds from overhunting. Since then, the refuge system has protected other birds. For example, in the 1930s, it was discovered that the

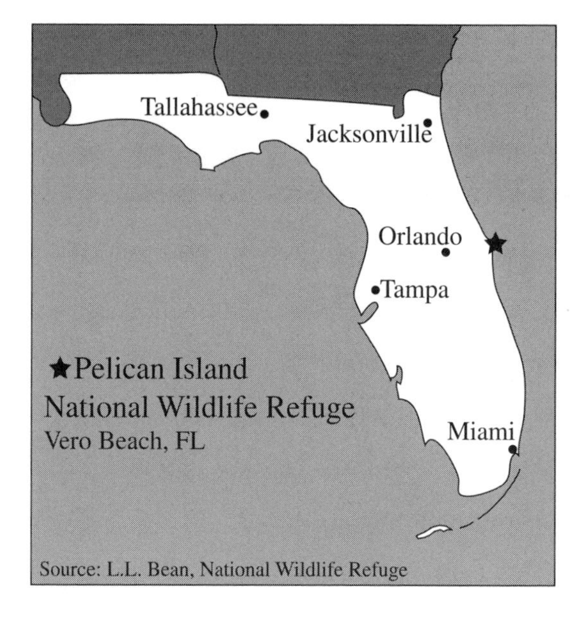

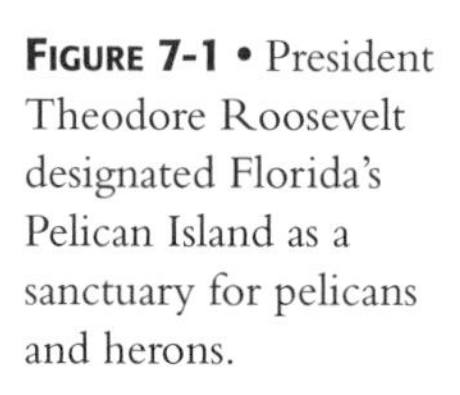

FIGURE 7-1 • President Theodore Roosevelt designated Florida's Pelican Island as a sanctuary for pelicans and herons.

whooping crane, a large and quite beautiful *migratory* bird, had been hunted to near extinction. Scientists had estimated that there were fewer than 50 birds left in both Canada and the United States. Then in 1937, the Arkansas National Wildlife Refuge was created in Texas to protect the wintering grounds of whooping cranes. Working with the Canadian government, wildlife managers were also able to establish protection for the birds on their breeding grounds in northern Canada and along their 4,000-kilometer (2,500-mile) migration route through North America. Although whooping cranes remain an *endangered species* in North America today, their population is slowly growing. Pelican Island is now part of the National Wilderness Preservation System and is the smallest of all wilderness areas.

Presently, the National Wildlife Refuge System embraces more than 500 such refuges on 37 million hectares (92 million acres) scattered across 50 states and several overseas territories. Refuges in the United States range in size from Minnesota's Mille Lacs (less than an acre) to Alaska's Yukon Delta at about 9 million hectares (20 million acres). The vast majority of these lands are located in Alaska, with the rest spread across the rest of the United States and several U.S. territories.

Many refuges permit hunting and fishing in season, as well as other recreational activities, such as hiking and swimming. As a result of these activities, critics fear some parts of the system have become too public.

Wilderness Conservation

Some refuges have been designated as wilderness areas. A wilderness is an undisturbed land area that receives protection under the Wilderness Act of 1964. Today nearly 500 wilderness areas, covering almost 38.5

TABLE 7-1 National Wildlife Refuges Established for Endangered Species

State	Unit Name	Species of Concern	Unit Acreage
Alabama	Blowing Wind Cave NWR	Indiana Bat, Gray Bat	264
Arkansas	Logan Cave NWR	Cave Crayfish, Gray Bat, Indiana Bat, Ozark Cavefish	124
Arizona	Buenos Aires NWR	Masked Bobwhite Quail	116,585
California	Bitter Creek NWR	California Condor	14,054
Florida	Archie Carr NWR	Loggerhead Sea Turtle, Green Sea Turtle	29
Hawaii	Hakalau Forest NWR	Hawaiian Hawk	32,730
Iowa	Driftless Area NWR	Iowa Pleistocence Snail	521
Massachusetts	Massasoit NWR	Plymouth Red-bellied Turtle	184
Missouri	Ozark Cavefish NWR	Ozark Cavefish	42
Nevada	Ash Meadows NWR	Devil's Hole Pupfish, Warm Springs Pupfish	13,268
Oklahoma	Ozark Plateau NWR	Ozark Big-eared Bat, Gray Bat	2,208
Texas	Attwater Prairie Chicken NWR	Attwater's Greater Prairie Chicken	8,007
Virgin Islands	Gray Cay NWR	St. Croix Ground Lizard	14
Washington	Julia Butler Hansen Refuge for Columbian White-tailed Deer	Columbian White-tailed Deer	2,777
Wyoming	Mortenson Lake NWR	Wyoming Toad	1,776

million hectares (95 million acres), exist throughout the United States. These wilderness areas provide habitat for countless species of wild animals and plants. Wilderness areas, along with national parks, national wildlife refuges, and national forests are public lands and can therefore be used by the public for a variety of purposes, including logging, mining, oil drilling, and agriculture, as well as for recreational activities such as boating, camping, and fishing.

The U.S. Department of the Interior, the National Park Service (NPS), the Fish and Wildlife Service (FWS), and the Bureau of Land Management (BLM) are the federal agencies responsible for overseeing the use of wilderness areas. Of all the public lands, wilderness areas are the most protected. Whereas national parks, forests, and wildlife refuges allow varying amounts of development such as logging, mining, oil drilling, and animal grazing, wilderness areas can be used only sparingly for these activities.

Before the Wilderness Act was passed, people argued about whether these delicate ecosystems should be preserved. Today the debate is often about how much land should be protected. Many people fear that "locking up" too much land as wilderness will be harmful to businesses and the economy. In addition, such restrictions will limit availability of land needed for the construction of new roads, bridges, and homes as the human population increases. More natural resources, such as minerals and timber, will also be needed to complete these projects. Others believe that more wilderness areas are needed to protect the country's many endangered and *threatened species*. They point out that even if it is illegal to trap, harm, or kill these species, they are still threatened indirectly from acid rain, water pollution, and other problems caused by human activities. Others also stress that compared to other public lands, relatively few places are even classified as wilderness areas. They argue that these remaining lands should be preserved for human enjoyment before they are gone forever.

Refer to Volume IV for more information about endangered and threatened species.

PARKS AND FORESTS

National Park System

The national park concept is generally credited to the artist and environmental activist George Catlin. On a trip to the Dakotas in 1832, Catlin was concerned about the impact of America's westward expansion on Indian civilization, wildlife, and wilderness. With the help of Catlin and others, the U.S. Congress in 1864 donated Yosemite Valley to California for preservation as a state park. Soon after, Congress approved the first national park, Yellowstone. After Yellowstone, other national parks were established in the 1890s and early 1900s, including Sequoia, Yosemite (to which California returned Yosemite Valley), Mount Rainier, Crater Lake, and Glacier.

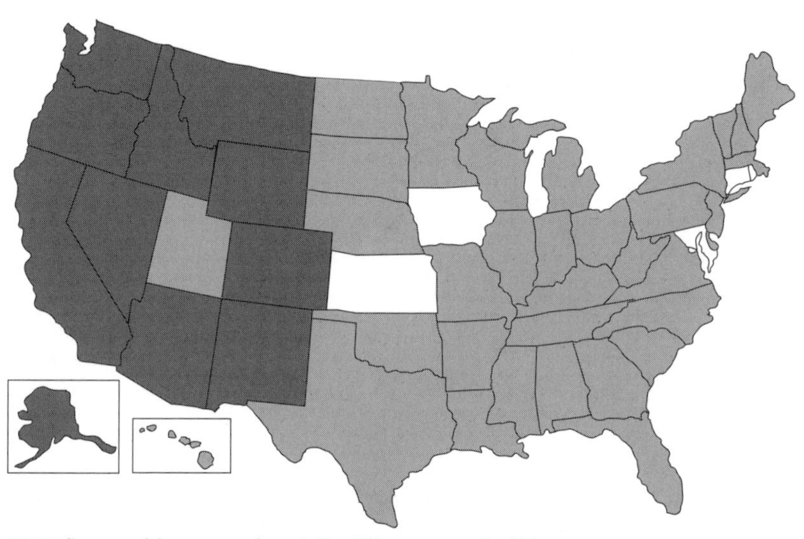

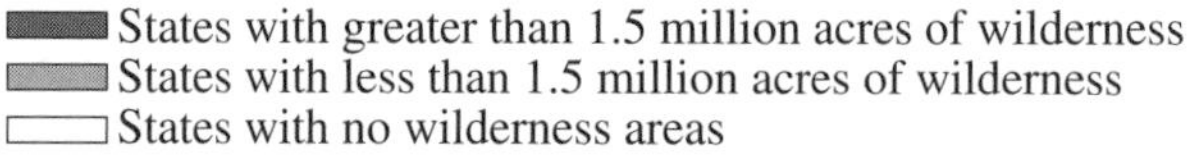

FIGURE 7-2 • National Park System

The giant sequoias in Yosemite National Park are some of the oldest and largest plants on Earth. (Courtesy of Emy Volpe)

The late nineteenth century also saw growing interest in preserving prehistoric Indian ruins and artifacts on the public lands. In 1906 Congress created Mesa Verde National Park, containing dramatic cliff dwellings in southwestern Colorado, and passed the Antiquities Act authorizing presidents to set aside "historic and prehistoric structures, and other objects of historic or scientific interest" in federal custody as national monuments.

Theodore Roosevelt added 18 national monuments before he left the presidency. They included not only cultural features like El Morro, New Mexico, the site of prehistoric *petroglyphs* and historic inscriptions, but natural features like Arizona's Petrified Forest and the Grand Canyon. Congress later converted many of these natural monuments to national parks. By 1916 the Interior Department was responsible for 14 national parks and 21 national monuments. In 1926 Congress authorized Shenandoah, Great Smoky Mountains, and Mammoth Cave as national parks in the Appalachian region. In 1937 Congress authorized Cape Hatteras National Seashore, the first of several seashore and lakeshore areas. The National Trails System Act of 1968 made the

El Capitan, the largest single granite rock on Earth, is a favorite site to see in Yosemite National Park in California. El Capitan rises about 1230 meters above sea level. (Courtesy of Emy Volpe)

National Park Service responsible for the Appalachian National Scenic Trail, running some 3,300 kilometers (2,000 miles) from Maine to Georgia. Gateway National Recreation Area in New York City and Golden Gate National Recreation Area in San Francisco were established in 1972.

As of 1999 the national park system comprised 379 areas in nearly every state and U.S. possession. In addition to managing these parks—as diverse and different as Hawaii Volcanoes National Park and the Statue of Liberty National Monument—the Park Service supports the preservation of natural and historic places and promotes outdoor recreation.

National Forest System

The National Forest System is administered by the U.S. Forest Service (USFS). This is largest agency of the U.S. Department of Agriculture. The Forest Service is responsible for the management, protection, and use of America's forests and range lands. The National Forest system has jurisdiction over national forests and grasslands. There are 155 national forests and 20 grasslands in 44 states, Puerto Rico, and the

The Grand Teton National Park is in Wyoming. (Courtesy of Peter Mongillo)

FIGURE 7-3 • Selected National Parks, National Monuments, and the National Seashores

Yellowstone National Park

The oldest national park in the world, created by the U.S. Congress in 1872, Yellowstone National Park is the largest national park in the lower 48 states, encompassing 898,714 hectares (3,472 square miles) of land in Wyoming, Montana, and Idaho. Ninety-nine percent of the park's area remains undeveloped, providing a wide range of habitats that support one of North America's most diverse large-mammal populations. Yellowstone may be one of the last wilderness ecosystems remaining in the world's temperate zones.

The natural features that initially led to preservation of Yellowstone as a national park were its *geothermal* features, the Grand Canyon of the Yellowstone River, and the Yellowstone Lake. There are nearly 10,000 thermal features in Yellowstone, including 200–250 active geysers (75 percent of the world's total), the most famous of which is Old Faithful. The park's thermal features are located in the only undisturbed geyser basins left in the world. The park also boasts one of the world's largest volcanic explosion craters, a caldera measuring 45 km by 75 km (28 mi. by 47 mi.).

Yellowstone is home to more than 300 species of animals: 60 different mammals (including elk, bighorn sheep, grizzly bears, and bobcats), 18 species of fishes, and more than 225 bird species. Bison are the largest mammals in the park, and the park preserves the only place in the lower 48 states where a population of wild bison has persisted since prehistoric times. As of 1902, fewer than 50 of these animals remained. Fearing extinction of the species, the park imported 21 privately owned bison to add stock to the wild herd; by 1996, Yellowstone's bison numbers had increased to about 3,500.

Two Yellowstone mammals, the wolf and the grizzly bear, are listed as endangered in the lower 48 states. As of April 1999, about 110 wolves, many of which are monitored by means of radio collars, inhabited the Yellowstone ecosystem thanks to a successful, through controversial, reintroduction program begun several years before. About 250 grizzly bears live in and around Yellowstone Park. As a result of cooperative efforts of many land management agencies, the population is now at the recovery target set for Yellowstone by the National Park Service.

FIGURE 7-4 • The peregrine falcon lives in Yellowstone National Park. The principal prey of peregrine falcons is other birds, specifically ducks.

FIGURE 7-5 • In the 1990s, the U.S. Fish and Wildlife Service directed a program to reintroduce gray wolves to Yellowstone National Park and central Idaho.

FIGURE 7-6 • The bison is a native species to North American prairies and is the largest mammal in Yellowstone National Park.

Virgin Islands, of which national forests comprise 77.3 million hectares, an area the size of Texas.

The Forest Service's mission is summarized as "Caring for the Land and Serving People." The agency manages natural resources under a multiple-use, sustained-yield concept to achieve quality land management and meet people's diverse needs. The management of resources under the best combination of uses benefits the American people, ensures the productivity of the land, and protects the environment.

Congress established the Forest Service in 1905 to provide water and timber for the nation's benefit. When the nation's resource needs changed, Congress directed the agency to manage national forests for additional multiple uses and benefits, as well as for the sustained yield of renewable resources such as water, wildlife, and recreation.

Mount Rushmore is a popular National Monument located in South Dakota. The images sculptured on the side of Mount Rushmore are Presidents, George Washington, Thomas Jefferson, Theodore Roosevelt, and Abraham Lincoln. (Courtesy of National Park Service)

U.S. FISH AND WILDLIFE SERVICE

The U.S. Fish and Wildlife Service (FWS) is an agency within the United States Department of Interior. The major objective of this agency is to conserve, protect, and enhance fish and wildlife species and their habitats. Its major areas of responsibility are migratory birds, endangered species, certain marine mammals, and freshwater and *anadromous* fish. FWS agents provide expert biological advice to other federal agencies, states, industry, and members of the public concerning the conservation of fish and wildlife habitat that may be affected by development activities.

Establishment of the U.S. Fish and Wildlife Service

The origin of the U.S. Fish and Wildlife Service dates back to 1871, when Congress established the U.S. Fish Commission to study and reverse the decline in the nation's food fish populations. In 1903 the Fish Commission was placed under the Department of Commerce and renamed the Bureau of Fisheries. Then in 1940 the Bureau of Fisheries and the Bureau of Biological Survey were transferred to the Department of the Interior. Further reorganization came in 1956 when the Fish and Wildlife Act created the U.S. Fish and Wildlife Service.

Endangered Species Act

Today the Fish and Wildlife Service maintains a headquarters in Washington, D.C., seven regional offices, and nearly 700 field units and installations, including national wildlife refuges, fish hatcheries, ecological field offices, and law enforcement offices. A major function of the agency is the identification and recovery of endangered species. The Service leads federals effort to protect and restore animals and plants that are in danger of extinction, both in the United States and worldwide. Using the best scientific information available, FWS agents identify species that appear to be endangered or threatened. The species that meet the criteria of the Endangered Species Act are placed on the Interior Department's official List of Endangered and Threatened Wildlife and Plants.

An endangered species list is a survey developed and maintained by the U.S. Fish and Wildlife Service of plants and animals that are deemed to be either endangered species or threatened species. For purposes of listing, endangered species are defined as those at immediate risk of extinction and unlikely survive without direct human intervention. Threatened species are those that are abundant in parts of their range but are declining in total numbers and thus face the risk of extinction in the foreseeable future. In recent years, the Service has placed increased emphasis on two provisions of the Endangered Species Act. These provisions are designed to avoid or resolve conflicts between private development projects and the protection of endangered species. The Service also consults with other federal agencies and renders opinions on the effects of proposed federal projects on endangered species,

For more information about endangered and threatened species, refer to Volume IV.

TABLE 7-2 Total Endangered Species in the United States and Worldwide

Group	Endangered		Threatened		Total
	U.S.	Foreign	U.S.	Foreign	Species
Mammals	61	251	8	16	336
Birds	75	178	15	6	274
Reptiles	14	65	21	14	114
Amphibians	9	8	8	1	26
Fishes	69	11	41	0	121
Clams	61	2	8	0	71
Snails	18	1	10	0	29
Insects	28	4	9	0	41
Arachnids	5	0	0	0	5
Crustaceans	17	0	3	0	20
Animal subtotal	357	520	123	37	1,037
Flowering plants	540	1	132	0	673
Conifers	2	0	1	2	5
Ferns and others	26	0	2	0	28
Plant subtotal	568	1	135	2	706
Grand total	925	521	258	39	1,743

Source: United States Fish and Wildlife Service, 1998.

FIGURE 7-7 • The snowy egret is protected by the Migratory Bird Treaty Act. The Migratory Bird Treaty Act was enacted in 1918 to protect most species of common wild birds of the United States and therefore prevent the extinction of some species.

recommending ways for developments to avoid harm to endangered species.

The Service also cooperates with state wildlife agencies and the Canadian Wildlife Service to regulate migratory bird hunting. Presently FWS agents are also increasing efforts to identify nongame bird species that may be declining and to undertake efforts to restore them. Through the Partners for Wildlife Program, the Service provide technical and financial assistance to private landowners wishing to restore wildlife habitat on their properties, primarily wetlands, riparian habitat, and native prairie. To date nearly 11,000 landowners have participated in Partners for Wildlife, restoring approximately 210,000 acres of habitat.

Restoring nationally significant fisheries that have been depleted by overfishing, pollution, or other habitat damage is also a responsibility of the U.S. Forest and Wildlife Service. Currently, FWS fishery specialists are devoting much of their efforts to helping four important fish groups: lake trout in the upper Great Lakes; striped bass of the Chesapeake Bay region and the Gulf Coast; Atlantic salmon of New England; and chinook, coho, and steelhead salmon of the Pacific Northwest. The FWS fishery program also works to compensate for losses of fishery resources caused by federal water projects and to improve fishery resources on Indian reservations and federal lands.

Two laws administered by the Service, the Federal Aid in Wildlife Restoration Act and the Federal Aid in Sport Fisheries Restoration Act, have created some of the most successful programs in the history of fish and wildlife conservation. These programs provide federal grant money to support specific projects carried out by state fish and wildlife agencies. The money comes from federal excise taxes on sporting arms and ammunition, archery equipment, and sportfishing tackle.

BUREAU OF LAND MANAGEMENT

The Bureau of Land Management (BLM) is an agency of the U.S. Department of the Interior that is responsible for managing 107 million hectares (264 million acres) of public land and resources. The area covers about one-eighth of the land in the United States and about 122 million additional hectares (300 million acres) of mineral estate. Public land is any land or interest in land be owned by the United States and administered by the Secretary of the Interior through the Bureau of Land Management. This would include public domain and acquired lands.

The Bureau of Land Management also is responsible for wildfire management on a total of 157 hectares (388 million acres). The resources on the public land include minerals, wild horses, *archeological* and historical artifacts, national monuments, and timber.

Most of the lands managed by the Bureau of Land Management are located in the western United States, including Alaska, and represent many ecosystems, including grasslands, forests, mountains, tundras, and deserts. The BLM manages a wide variety of resources and uses, including energy and minerals; timber; fish and wildlife habitat; wilderness areas; and natural and cultural heritage resources.

BLM origins date to the Land Ordinance of 1785 and the Northwest Ordinance of 1787, which provided for the survey and settlement of lands ceded to the federal government by the 13 original colonies following the Revolutionary War. As the United States acquired additional lands from other countries, the Congress directed that they be explored, surveyed, and made available for settlement. In 1812, Congress

FIGURE 7-8 • A Selection of National Monuments in Southwestern United States

The Statue of Liberty is a National Monument. (Courtesy of Statue of Liberty National Monument)

established the General Land Office within the Department of the Treasury to oversee the disposition of these lands and to encourage settlement of the land by enacting laws such as the Homestead Act of 1862 and the General Mining Law of 1872.

In the late nineteenth century, new federal land management priorities created the first national parks, national forests, and wildlife refuges. By withdrawing these lands from settlement, Congress formalized a shift in the purposes that public lands were expected to serve. There was a new recognition that such lands should be held in public ownership because of notable natural resource values. In the early twentieth century, Congress took additional steps toward recognizing these values. For example, the Mineral Leasing Act of 1920 permitted leasing, exploration, and production of certain commodities, such as coal, petroleum, natural gas, and sodium. The Taylor Grazing Act of 1934 established the

Rainbow Bridge National Monument is located in Utah. (Courtesy of United States Geological Survey)

Muir Woods is a National Monument located in Mill Valley, California. Giant redwood trees dominate the 300 acre park. (Courtesy of John D. Mongillo)

U.S. Grazing Service to manage the public range lands; the Oregon and California (O&C) Act of August 28, 1937, required *sustained-yield management* of the forests of western Oregon.

RIVER SYSTEMS AND NATIONAL TRAILS

The National Wild and Scenic Rivers System and the National Trails System are both administered by the Forest Service, the Bureau of Land Management, and the National Park Service to manage the nation's waterways and their environs and to oversee identified networks of trails.

Wild and Scenic Rivers Act

The Wild and Scenic Rivers Act was enacted in 1968 to protect certain rivers from being dammed or damaged in other ways. As part of the Wild and Scenic Rivers Act, federal land management agencies, including the U.S. Forest Service and the National Park Service, were directed to identify rivers that could be included in a National Wild and Scenic Rivers System. Rivers and river segments protected under this law fall into one of three categories: wild, scenic, or recreational. Wild rivers are undisturbed rivers that are accessible only by trails. Scenic rivers are mostly untouched, but they are accessible by road.

TABLE 7-3 National Wild and Scenic Rivers System (2003)

River	Administering Agency	Miles by Classfication	
		Wild	*Scenic*
Rio Grande, New Mexico	Forest Service and Bureau of Land Management	53.2	—
Rio Grande, Texas	National Park Service	95.2	96.0
St. Croix, Minnesota & Wisconsin	National Park Service	—	181.0
Salmon, Idaho	Forest Service	79.0	—
Allagash Wilderness Waterway, Maine	State of Maine	92.5	—
Snake, Idaho & Oregon	Forest Service	32.5	34.4
Missouri, Montana	Bureau of Land Management	64.0	26.0
Obed, Tennessee	National Park Service	43.4	2.0
Delaware (Upper), New York & Pennsylvania	National Park Service	—	23.1
Charley, Alaska	National Park Service	208.0	—
Fortymile, Alaska	Bureau of Land Management	179.0	203.0
Verde, Arizona	Forest Service	22.2	18.3
Merced, California	Forest Service	15.0	2.0
	National Park Service	53.0	14.0
	Bureau of Land Management	3.0	—
Sipsey Fork West Fork, Alabama	Forest Service	36.4	25.0
North Fork John Day, Oregon	Forest Service	27.8	10.5
Ontonagon, Michigan	Forest Service	42.9	41.0
Sudbury, Assabet, Concord, Massachusetts	National Park Service State of Massachusetts and Local Government	—	14.9

Zoos

Zoos have also become an important site for the preservation and protection of wildlife resources, particularly those species that are endangered. At one time, many zoos displayed live animals for public entertainment. Presently some zoos have become scientific and educational institutions that have contributed to the understanding and conservation of wild animal populations. Among challenges facing modern zoos are the costs of upgrading old facilities, the struggle to obtain sufficient operating funds, and the need to attract more visitors in new and entertaining exhibits.

Many older zoos in American cities have undergone renovations during the last decades of the twentieth century. Among the recent trends in zoo improvement is the construction of new enclosures that resemble natural habitats. The replacement of traditional steel bars and sawdust-covered concrete floors with appropriately designed surroundings improves visitor appreciation of the animals. Such renovations may reduce stress on animals and allow them to interact with one another more naturally.

Several major zoos conduct captive propagation programs. A captive propagation program includes the breeding of selected zoo or wild animals to obtain offspring, usually for release to the wild or for transfer to other zoos. Captive breeding is one method of saving some species from extinction. The captive breeding program at the San Diego Zoo, for example, produced California condors (an endangered species) that were later successfully released to the wild. Some zoos have extended their activities to conservation programs in other countries, assisting in the establishment of wildlife preserves.

Zoos have expanded and improved public education programs also, with education departments that develop programs related to zoo exhibits. Public outreach activities include in-school programs, zoo tours, special events, and Websites. The Zoological Society of New York, for example, conducted a joint project with the Cameroon government of West Africa to monitor an elephant herd as it moved throughout its range.

In the view of critics, many remaining older zoos provide poor living conditions for the animals they keep. Visitors to these zoos may see old, unhealthy, or dysfunctional wildlife specimens. Unfortunately, some zoos have acted irresponsibly in the acceptance of threatened or endangered wildlife species or animals collected by illegal means, such as poaching. Opponents of such zoos believe that the zoos seek to profit from exploitation of animals, with the effect of rewarding poachers. Often, however, zoos provide a sanctuary for animals seized from illegal wildlife traffickers, circuses, or illicit traveling zoos that improperly handle live animals.

The importance of zoos will increase as natural habitats dwindle. Through their efforts in support of conservation, education, and environmental advocacy, zoos will continue to play a critical role in wildlife preservation throughout the world.

Polar bears also live in zoos. An adult polar bear may weigh as much as 450 kilograms. These two live in the Roger Williams Zoo in Providence, Rhode Island. (Courtesy of Roger Williams Park Zoo, Providence, Rhode Island)

Snow leopards have been hunted for their skins and are on the endangered species list. Some rare snow leopards are found in zoos. This one lives at the Roger Williams Zoo in Providence, Rhode Island. (Courtesy of Roger Williams Park Zoo, Providence, Rhode Island)

Recreational rivers are those that are used for recreational activities such as swimming, fishing, and boating.

Today there are 152 designated rivers and river segments within the National Wild and Scenic Rivers System. Some of the rivers included are the American and Klamath rivers in California; the Delaware River in New York, New Jersey, and Pennsylvania; and the Rio Grande in Texas. The National Wild and Scenic Rivers Act strictly prohibits the construction of dams, hydroelectric power plants, and other structures on and around these rivers. The act also limits the extraction of minerals from any river within the national system.

The National Wild and Scenic Rivers Act has done much to protect the environment and beautify the land. Environmentalists, however, fear that the Wild and Scenic Rivers Act is much too flexible to be effective. Critics often point out that even when a river is included in the national system, any preexisting land use along that river, such as mining, logging, and farming, is permitted to continue.

National Trails System

In the United States there are several networks of trails that are administered by the Forest Service, the Bureau of Land Management, and the National Park Service. One of the best known is the Appalachian Trail, which was completed in 1937. Another trail system includes the Continental Divide Trail which covers the Rocky Mountains from Mexico to Canada. It is the most rugged of the long-distance trails. It is open to hikers, to pack and saddle animals, and in some places, to off-road vehicles. The National Trails System does not receive the same kind of protection as wildlife refuges and parks. However, there are several volunteer groups, clubs, and organizations that assist the government agencies by upkeeping and protecting trails. These groups also provide educational programs, safety skills on hiking, and search-and-rescue assistance.

Appalachian Trail

The Appalachian Trail is officially called the Appalachian National Scenic Trail. The hiking path passes through 14 states in the eastern United States. Conceived in 1921 by Benton MacKaye, a forester and regional planner, and completed in 1937, the trail extends along the ridges of the Appalachian Mountains from Mt. Katahdin, Maine, to Springer Mt. Georgia. The largest part of the trail passes through eight national forests and two national parks, but some of its length is still on private property. Hiking and trail clubs maintain shelters and campsites along the path. The Appalachian and Pacific Crest scenic trails, both designated in 1968, were the first two components in the National Trails System.

Vocabulary

Anadromous Describes fish species that hatch in freshwater and live as an adult in salt water.

Archeological Related to the scientific study of the life and culture of ancient people.

Endangered species A specific plant or animal species whose population is so small in numbers that the species is in danger of disappearing.

Geothermal Describes hot water or steam produced by the transfer of heat from Earth's interior.

Migratory Describes an animal such as a bird that moves from one location to another.

Petroglyph A prehistoric rock carving.

Sanctuary A special area where wildlife is protected.

Sustained-yield management A harvest level for natural resources, such as timber or crops, during a given time period that can be continued over time without jeopardizing the ability of the environment to recover.

Threatened species A specific plant or animal species whose population level in some sections of its natural range is very low.

Wildlife Under federal regulations, living or dead undomesticated animals, even if bred, hatched, or born in captivity, including any part, product, egg, or offspring of such animals. The term applies to wild mammals, birds, fish, reptiles, amphibians, and other terrestrial or aquatic life.

Activities for Students

1. Contact your local Audubon and wilderness societies as well as your local zoos and park systems. Determine what wild life is endangered in your area. What have been the major causes of this, and what are the local agencies doing to help the situation?
2. What was happening politically at the turn of the century to inspire people, under Theodore Roosevelt, to preserve the parks and wildlife?

Books and Other Reading Materials

Albright, Horace M., and Marian Albright Schenck. *Creating the National Park Service: The Missing Years.* Norman: University of Oklahoma Press, 1999.

Doppelt, Bob, Mary Scurlock, Chris Frissell, and James Karr. *Entering the Watershed: A New Approach to Save America's River Ecosystems.* Washington, D.C.: Island Press. 1993.

Hartzog, George B., Jr. *Battling for the National Parks.* South Kingston, R.I.: Moyer Bell, 1988.

Mackintosh, Barry. *The National Parks: Shaping the System.* Washington D.C.: National Park Service, 1991.

National Geographic, ed. *National Geographics Guide to the National Parks of the United States.* Washington, D.C.: National Geographic Society, 2001.

Oelschlaeger, Max. *The Idea of Wilderness: From Prehistory to the Age of Ecology.* New Haven, Conn.: Yale University Press, 1993.

Palmer, Tim. *The Wild and Scenic Rivers of America.* Washington, D.C.: Island Press, 1993.

Rettie, Dwight F. *Our National Park System: Caring for America's Greatest Natural and Historic Treasures.* Urbana: University of Illinois Press, 1995.

Sellars, Richard West. *Preserving Nature in the National Parks: A History.* New Haven, Conn.: Yale University Press, 1997.

Shankland, Robert. *Steve Mather of the National Parks. 3d ed.* New York: Alfred A. Knopf, 1976.

Websites

Bronx Zoo, http://www.bronxzoo.com/

Bureau of Land Management, http://www.blm.gov

Conservation International, http://www.conservation.org

Convention on International Trade in Endangered Species (CITES) homepage, http://www.cites.org

IUCN, World Conservation Union/International Union for the Conservation of Nature, http://www.iucn.org

National Audubon Society, http://www.audubon. org

National Fish and Wildlife Foundation, http://www.nfwf.org

National Forest Foundation, http://www.nffweb.org

National Park Service, Yellowstone National Park, http://www.nps.gov/yell

National Wildlife Refuge System, http://www.refuges.fws.gov/NWRSHomePage.html

The Nature Conservancy, http://www.tnc.org

Public Lands Foundation, http://www.publicland.org

San Diego Zoo, http://www.sandiegozoo.org/

U.S. Fish and Wildlife Service, U.S. Department of the Interior, http://www.fws.gov

U.S. Forest Service, http://www.fs.fed.us

Yellowstone Ecosystem Studies, http://www.yellowstone.org

Yellowstone Net, http://www.yellowstone.net

Appendix A: Environmental Timeline, 1620–2004

Environmentalists and activists appear in **boldface**.

1620 to 1860 Erosion becomes a major problem on many American farms. Fields are abandoned. Rivers and streams are filled with silt and mud. The publication of farm journals is initiated by early soil conservationists to improve farming methods.

1748 Jared Eliot, a minister and doctor of Killingsworth, Connecticut, writes the first American book on agriculture to improve crops and to conserve soil.

1824 Solomon and William Drown of Providence, Rhode Island, publish *Farmer's Guide* which discusses erosion and its causes and remedies. A year later, John Lorain, of the Philadelphia Agricultural Society, publishes a book devoted to the prevention of soil erosion in which he discusses methods such as using grass as an erosion-control crop.

1827 John James Audubon begins publication of *Birds of America*.

1830 George Catlin launches his great western painting crusade to document Native American peoples.

1845 Henry David Thoreau moves to Walden Pond to observe the fauna and flora of Concord, Massachusetts.

1847 U.S. Congressman **George Perkins Marsh** of Vermont delivers a speech calling attention to the destructive impact of human activity on the land.

1849 The U.S. Department of the Interior (DOI) is established.

1857 Frederick Law Olmsted develops the first city park: New York City's Central Park.

1859 British naturalist Charles Darwin publishes *The Origin of the Species by Means of Natural Selection*. In time the theory of evolution presented in the book becomes the most widely accepted theory of evolution.

1866 German biologist Ernst Haeckel introduces the term *ecology*.

1869 John Muir moves to the Yosemite Valley.

Geologist and explorer John Wesley Powell travels the Colorado River through the Grand Canyon.

1872 Yellowstone National Park is established as the first national park of the United States in Yellowstone, Wyoming.

U.S. legislation: Passage of the Mining Law permits individuals to purchase rights to mine public lands.

1876 The Appalachian Mountain Club is founded.

1879 The U.S. Geological Survey (USGS) is formed.

1882 The first hydroelectric plant opens on the Fox River in Wisconsin.

1883 Krakatoa, a small island of Indonesia, is virtually destroyed by a volcanic explosion.

1890 Denmark constructs the first windmill for use in generating electricity.

Sequoia National Park, Yosemite National Park, and General Grant National Park are established in California.

1891 U.S. legislation: Passage of Forest Reserve Act provides the basis for a system of national forests.

1892 John Muir, Robert Underwood Johnson, and William Colby are cofounders of the Sierra Club, in Muir's words, to "do something for wildness and make the mountains glad."

1893 The National Trust is founded in the United Kingdom. The group purchases land deemed of having natural beauty or considered a cultural landmark.

1895 Founding of the American Scenic and Historic Preservation Society.

1898 Cornell University establishes the first college program in forestry.

Gifford Pinchot becomes head of the U.S. Division of Forestry (now the U.S. Forest Service) and serves until 1910. Under President

Theodore Roosevelt, many of Pinchot's ideas became national policy. During his service, the national forests increase from 32 in 1898 to 149 in 1910, a total of 193 million acres.

1899 The River and Harbor Act bans pollution of all navigable waterways. Under the act, the building of any wharves, piers, jetties, and other structures is prohibited without congressional approval.

1900 U.S. legislation: Passage of Lacey Act makes it unlawful to transport illegally killed game animals across state boundaries.

1902 U.S. legislation: Passage of Reclamation Act establishes the Bureau of Reclamation.

1903 First federal U.S. wildlife refuge is established on Pelican Island in Florida.

1905 The National Audubon Society, named for wildlife artist John James Audubon, is founded.

1906 Yosemite Valley is incorporated into Yosemite National Park.

1907 International Association for the Prevention of Smoke is founded. The group's name later changes several times to reflect other concerns over causes of air pollution.

Gifford Pinchot is appointed the first chief of the U.S. Forest Service.

1908 The Grand Canyon is set aside as a national monument.

Chlorination is first used at U.S. water treatment plants.

President Theodore Roosevelt hosts the first Governors' Conference on Conservation.

1914 The last passenger pigeon, Martha, dies in the Cincinnati zoo.

1916 The National Park Service (NPS) is established.

1918 Hunting of migratory bird species is restricted through passage of the Migratory Bird Treaty Act. The act supports treaties between the United States and surrounding nations.

Save-the-Redwoods League is created.

1920 U.S. legislation: Passage of the Mineral Leasing Act regulates mining on federal lands.

1922 The Izaak Walton League is organized under the direction of **Will H. Dilg**.

1924 Environmentalist **Aldo Leopold** wins designation of Gila National Forest, New Mexico, as first extensive wilderness area.

Marjory Stoneman Douglas, of the *Miami Herald*, writes newspaper columns opposing the draining of the Florida Everglades.

Bryce Canyon National Park is established in Utah.

1925 The Geneva Protocol is signed by numerous countries as a means of stopping use of biological weapons.

1928 The Boulder Canyon Project (Hoover Dam) is authorized to provide irrigation, electric power, and a flood-control system for Arizona and Nevada communities.

1930 Chlorofluorocarbons (CFCs) are deemed safe for use in refrigerators and air conditioners.

1931 France builds and makes use of the first Darrieus aerogenerator to produce electricity from wind energy.

Addo Elephant National Park is established in the Eastern Cape region of South Africa to provide a protected habitat for African elephants.

1932 Hugh Bennett is given the opportunity to put his soil conservation ideas into practice to help reduce soil erosion. He becomes the director of the Soil Erosion Service (SES) created by the Department of Interior.

1933 The Tennessee Valley Authority (TVA) is formed.

The Civilian Conservation Corps (CCC) employs more than 2 million Americans in forestry, flood control, soil erosion, and beautification projects.

1934 The greatest drought in U.S. history continues. Portions of Texas, Oklahoma, Arkansas, and several other midwestern states are known as the "Dust Bowl."

U.S. legislation: Passage of Taylor Grazing Act regulates livestock grazing on federal lands.

1935 The Soil Conservation Service (SCS) is established.

The Wilderness Society is founded.

1936 The National Wildlife Federation (NWF) is formed.

1939 David Brower produces his first nature film for the Sierra Club, called *Sky Land Trails of the Kings*. In the same year, Brower, who is an excellent climber, completes his most famous ascent, Shiprock, a volcanic plug which rises 1,400 feet from the floor of the New Mexico desert.

1940 The U.S. Wildlife Service is established to protect fish and wildlife.

U.S. legislation: President Franklin Roosevelt signs the Bald Eagle Protection Act.

1945 The United Nations (UN) establishes the Food and Agriculture Organization (FAO).

1946 The International Whaling Commission (IWC) is formed to research whale populations.

The U.S. Bureau of Land Management (BLM) and the Atomic Energy Commission (AEC) are created.

1947 Marjory Stoneman Douglas publishes *The Everglades: River of Grass* and serves as a member of the committee that gets the Everglades designated a national park.

1948 The UN creates the International Union for the Conservation of Nature (IUCN) as a special environmental agency.

An air pollution incident in Donora, Pennsylvania, kills 20 people; 14,000 become ill.

U.S. legislation: Passage of Federal Water Pollution Control Law.

1949 Aldo Leopold's *A Sand County Almanac* is published posthumously.

1950 Oceanographer **Jacques Cousteau** purchases and transforms a former minesweeper, the *Calypso*, into a research vessel which he uses to increase awareness of the ocean environment.

1951 Tanzania begins its national park system with the establishment of the Serengeti National Park.

1952 Clean air legislation is enacted in Great Britain after air pollution–induced smog brings about the deaths of nearly 4,000 people.

David Brower becomes the first executive director of the Sierra Club.

1953 Radioactive iodine from atomic bomb testing is found in the thyroid glands of children living in Utah.

1955 U.S. legislation: Passage of the Air Pollution Control Act, the first federal legislation designed to control air pollution.

1956 U.S. legislation: Passage of the Water Pollution Control Act authorizes development of water-treatment plants.

1959 The Antarctic Treaty is signed to preserve natural resources of the continent.

1961 The African Wildlife Foundation (AWF) is established as an international organization to protect African wildlife.

1962 **Rachel Carson** publishes *Silent Spring*, a groundbreaking study of the dangers of DDT and other insecticides.

Hazel Wolf joins the National Audubon Society in Seattle, Washington, and plays a prominent role in local, national, and international environmental efforts during her lifetime.

1963 The Nuclear Test Ban Treaty between the United States and the Soviet Union stops atmospheric testing of nuclear weapons.

U.S. legislation: Passage of the first Clean Air Act (CAA) authorizes money for air pollution control efforts.

1964 **Hazel Henderson** organizes women in a local play park in New York City and starts a group called Citizens for Clean Air, the first environmental group, she believes, east of the Mississippi. She built Citizens for Clean Air from a very small group to a membership of 40,000. Two years later, 80 people died in New York City from air pollution–related causes during four days of atmospheric inversion.

U.S. legislation: Passage of the Wilderness Act creates the National Wilderness Preservation System.

1965 U.S. legislation: Passage of the Water Quality Act authorizes the federal government to set water standards in absence of state action.

1966 Eighty people in New York City die from air pollution–related causes.

1967 The *Torey Canyon* runs aground spilling 175 tons of crude oil off Cornwall, England.

Dian Fossey establishes the Karisoke Research Center in the Virunga Mountains, within the Parc National des Volcans in Rwanda to study endangered mountain gorillas.

The Environmental Defense Fund (EDF) is formed to lead an effort to save the osprey from DDT.

1968 U.S. legislation: Passage of the Wild and Scenic Rivers Act and the National Trails System Act identify areas of great scenic beauty for preservation and recreation.

Paul Ehrlich publishes *The Population Bomb*.

1969 Wildlife photographer Joy Adamson establishes the Elsa Wild Animal Appeal, an organization dedicated to the preservation and humane treatment of wild and captive animals.

Greenpeace is created.

Blowout of oil well in Santa Barbara, California, releases 2,700 tons of crude oil into the Pacific Ocean.

U.S. legislation: Passage of the National Environmental Policy Act (NEPA) requires all federal agencies to complete an environmental impact statement for any dam, highway, or other large construction project undertaken, regulated, or funded by the federal government.

The Friends of the Earth (FOE) is founded in the United States.

John Todd, **Nancy Jack Todd**, and Bill McLarney are the cofounders of the New Alchemy Institute in Cape Cod, Massachusetts. The institute begins to pioneer a new way of treating sewage and other wastes.

1970 **Denis Hayes** is the national coordinator of the first Earth Day, which is celebrated on April 22.

Construction of the Aswan High Dam on the Nile River in Egypt is completed.

U.S. legislation: Passage of an amended Clean Air Act (CAA) expands air pollution control.

The U.S. Environmental Protection Agency (EPA) is established.

1971 Canadian primatologist Biruté Galdikas begins her studies of orangutans through the Orangutan Research and Conservation Project in Borneo.

The United Nations Educational, Scientific and Cultural Organization (UNESCO) establishes the Man and the Biosphere Program, developing a global network of biosphere reserves.

1972 The Biological and Toxin Weapons Convention is adopted by 140 nations to stop the use of biological weapons.

The EPA phases out the use of DDT in the United States to protect several species of predatory birds. The ban builds on information obtained from Rachel Carson's 1962 book, *Silent Spring*.

U.S. legislation: Passage of the Water Pollution Control Act, the Coastal Zone Management Act (CZMA), and the Environmental Pesticide Control Act.

Oregon passes the first bottle-recycling law.

1973 Norwegian philosopher Arne Naess coins the term *deep ecology* to describe his belief that humans need to recognize natural things for their intrinsic value, rather than just for their value to humans.

The Convention on International Trade in Endangered Species of Wild Fauna and Flora (CITES) is signed by more than 80 nations. The Endangered Species Act of the United States also is enacted.

Congress approves construction of the 1,300-kilometer pipeline from Alaska's North Slope oil field to the Port of Valdez.

An Energy crisis in the United States arises from an Arab oil embargo.

A collision and resulting explosion between the *Corinthos* oil tanker and the *Edgar M. Queeny* releases 272,000 barrels of crude oil and other chemicals into the Delaware River near Marcus Hook, Pennsylvania.

1974 Scientists report their discovery of a hole in the ozone layer above Antarctica.

U.S. legislation: Passage of the Safe Drinking Water Act sets standards to protect the nation's drinking water. The EPA bans most uses for aldrin and dieldrin and disallows the production and importation of these chemicals into the United States.

1975 Unleaded gas goes on sale. New cars are equipped with antipollution catalytic converters.

The EPA bans use of asbestos insulation in new buildings.

Edward Abbey publishes *The Monkey Wrench Gang*, a novel detailing acts of ecotage as a means of protecting the environment.

1976 *Argo Merchant* runs aground releasing 25,000 tons of fuel into the Atlantic Ocean near Nantucket, Rhode Island.

National Academy of Sciences reports that CFC gases from spray cans are damaging the ozone layer.

U.S. legislation: Passage of the Resource Conservation and Recovery Act empowers the EPA to regulate the disposal and treatment of municipal solid and hazardous wastes. The Toxic Substances Control Act and the Resource Conservation and Recovery Act are enacted.

Fire aboard the *Hawaiian Patriot* releases nearly 100,000 tons of crude oil into the Pacific Ocean.

1977 The Green Belt Movement is begun by Kenyan conservationist Wangari Muta Maathai on World Environment Day.

Blowout of Ekofisk oil well releases 27,000 tons of crude oil into the North Sea.

Construction of the Alaska pipeline, the 1,300-kilometer pipeline that carries oil from

Alaska's North Slope oil field to the Port of Valdez, is completed at a cost of more than $8 billion.

U.S. legislation: Passage of the Surface Mining Control and Reclamation Act.

The Department of Energy (DOE) is created.

1978 The *Amoco Cadiz* tanker runs aground spilling 226,000 tons of oil into the ocean near Portsall, Brittany.

People living in the Love Canal community of New York are evacuated from the area to reduce their exposure to chemical wastes which have surfaced from a canal formerly used as a dump site.

Rainfall in Wheeling, West Virginia, is measured at a pH of 2, the most acidic rain yet recorded.

Aerosols with fluorocarbons are banned in the United States.

The EPA bans the use of asbestos in insulation, fireproofing, or decorative materials.

1979 British scientist **James E. Lovelock** publishes *Gaia: A New Look at Life on Earth.*

Collision of the *Atlantic Empress* and the *Aegean Captain* releases 370,000 tons of oil into the Caribbean Sea.

The Convention on Long-Range Transboundary Air Pollution (LRTAP) is signed by several European nations to limit sulfur dioxide emissions which cause acid rain problems in other countries.

The Three Mile Island Nuclear Power Plant in Pennsylvania experiences near-meltdown.

The EPA begins a program to assist states in removing flaking asbestos insulation from pipes and ceilings in school buildings throughout the United States.

The EPA bans the marketing of herbicide Agent Orange in the United States.

1980 Debt-for-nature swap idea is proposed by Thomas E. Lovejoy: nations could convert debt to cash which would then be used to purchase parcels of tropical rain forest to be managed by local conservation groups.

Global Report to the President addresses world trends in population growth, natural resource use, and the environment by the end of the century, and calls for international cooperation in solving problems.

U.S. legislation: Passage of the Comprehensive Environmental Response, Compensation, and Liability Act (Superfund) and the Low Level Radioactive Waste Policy Act.

1981 Earth First!, a radical environmental action group that resorts to ecotage to gain its objectives, formed.

Lois Gibbs founds the Citizens' Clearinghouse for Hazardous Wastes, later named the Center for Health, Environment, and Justice (CHEJ).

1982 U.S. legislation: Passage of the Nuclear Waste Policy Act.

1983 A film of **Randy Hayes**, *The Four Corners, a National Sacrifice Area*, wins the 1983 Student Academy Award for the best documentary. The film documents the tragic effects of uranium and coal mining on Hopi and Navajo Indian lands in the American Southwest.

The residents of Times Beach, Missouri, are ordered to evacuate their community. Investigations of Times Beach in the 1980s disclosed the fact that oil contaminated with dioxin, a highly toxic substance, had been used to treat the town's streets.

Cathrine Sneed founds and acts as director of the Garden Project in San Francisco. The Garden Project, a horticulture class for inmates of the San Francisco County Jail, uses organic gardening as a metaphor for life change. The U.S. Department of Agriculture calls the project "one of the most innovative and successful community-based crime prevention programs in the country."

1984 Toxic gases released from the Union Carbide chemical manufacturing plant in Bhopal, India kill an estimated 3,000 people and injure thousands of others.

The Jane Goodall Institute (JGI) is founded.

The British tanker *Alvenus* spills 0.8 million gallons of oil into the Gulf of Mexico.

U.S. legislation: Passage of the Hazardous and Solid Waste Amendments.

1985 Concerned Citizens of South Central Los Angeles becomes one of the first African American environmental groups in the United States. **Julia Tate** serves as the executive director. The organization's goal is to provide a better quality of life for the residents of this Los Angeles community. **Maria Perez**, **Nevada Dove**, and **Fabiola Tostado** later join the group and are known as the Toxic Crusaders.

Huey D. Johnson becomes the founder and president of the Resource Renewal Institute

(RRI), a nonprofit organization located in California. Johnson suggests that green plans is the path countries should take to respond to environmental decline. Green plans treat the environment as it really exists—a single, interconnected ecosystem that can be safeguarded for future generations only through a systemic, long-range plan of action.

Scientists of the British Antarctica Survey discover the ozone hole. The hole, which appears during the Antarctic spring, is caused by the chlorine from CFCs.

Juana Gutiérrez becomes president and founder of Mothers of East Los Angeles, Santa Isabel Chapter (Madres del Este de Los Angeles—Santa Isabel) (MELASI) whose mission is to fight against toxic dumps and incinerators and also to take a proactive approach to community improvement.

Primatologist Dian Fossey is discovered murdered in her cabin at the Karosoke Research Center she founded. Her death is attributed to poachers.

While protesting nuclear testing being conducted by France in the Pacific Ocean, the *Rainbow Warrior* (a boat owned by Greenpeace) is sunk in a New Zealand harbor by agents of the French government.

U.S. legislation: Passage of the Food Security Act.

1986 Tons of toxic chemicals stored in a warehouse owned by the Sandoz pharmaceutical company are released into the Rhine River near Basel, Switzerland. The effects of the spill are experienced in Switzerland, France, Germany, and Luxembourg.

An explosion destroys a nuclear power plant in Chernobyl, Ukraine, immediately killing more than 30 people and leading to the permanent evacuations of more than 100,000 others.

Bovine spongiform encephalopathy (BSE), a neurodegenerative illness of cattle, also known as mad cow disease, comes to the attention of the scientific community when it appears in cattle in the United Kingdom.

U.S. legislation: Passage of the Emergency Response and Community Right-to-Know Act and the Superfund Amendments and Reauthorization Act (SARA).

1987 The Montreal Protocol, an international treaty that proposes to cut in half the production and use of CFCs, is approved by more than 30 nations.

The world's fourth largest lake, the Aral Sea of Asia, is divided in two as a result of the diversion of water from its feeder streams, the Syr Darya and Amu Darya rivers.

The *Mobro*, a garbage barge from Long Island, New York, travels 9,600 kilometers in search of a place to offload the garbage it carries.

1988 Use of ruminant proteins in the preparation of cattle feed is banned in the United Kingdom to prevent outbreaks of BSE.

Global temperatures reach their highest levels in 130 years.

The Ocean Dumping Ban legislates international dumping of wastes in the ocean.

EPA studies report that indoor air can be 100 times as polluted as outdoor air. Radon is found to be widespread in U.S. homes.

Beaches on the east coast of the United States are closed because of contamination by medical waste washed onshore.

The United States experiences its worst drought in 50 years.

Plastic ring six-pack holders are required to be made degradable.

U.S. legislation: Passage of the Plastic Pollution Research and Control Act bans ocean dumping of plastic materials.

1989 The United Kingdom bans the use of cattle brains, spinal cords, tonsils, thymuses, spleens, and intestines in foods intended for human consumption as a means of preventing further outbreaks of Creutzfeldt-Jakob disease (CJD), the human version of mad cow disease, in humans.

Fire aboard the *Kharg 5* releases 75,000 tons of oil into the sea surrounding the Canary Islands.

The Montreal Protocol treaty is updated and amended.

The New York Department of Environmental Conservation reports that 25 percent of the lakes and ponds in the Adirondacks are too acidic to support fish.

The *Exxon Valdez* runs aground on Prince William Sound, Alaska, spilling 11 million gallons of oil into one of the world's most fragile ecosystems.

1990 **Ocean Robbins**, age 16, and **Ryan Eliason**, 18, are the cofounders of YES!, or Youth for Environmental Sanity. The goal of YES! is to educate, inspire, and empower young people to take positive action for healthy people and a healthy planet. Robbins served as director for five years and is now the organization's president. As of 2000, the program has reached 600,000 students in 1,200 schools in 43 states through full school assemblies.

UN report forecasts a world temperature increase of 2°F within 35 years as a result of greenhouse gas emissions.

U.S. legislation: Passage of the Clean Air Act amendments including requirements to control the emission of sulfur dioxide and nitrogen oxides.

1991 The Gulf War concludes with hundreds of oil wells in Kuwait being set afire by Iraqi troops, resulting in extensive air and water pollution problems.

The United States accepts an agreement on Antarctica which prohibits activities relating to mining, protects native species of flora and fauna, and limits tourism and marine pollution.

Eight scientists begin a two-year stay in Biosphere 2 in Arizona, a test center designed to provide a self-sustaining habitat modeling Earth's natural environments. The experiment, which is repeated in 1993, meets with much criticism and is deemed largely unsuccessful.

1992 UN Earth Summit is held in Rio de Janeiro, Brazil. Major resolutions resulting from the summit include the Rio Declaration on Environment and Development, Agenda 21, Biodiversity Convention, Statement of Forest Principles, and the Global Warming Convention, which is signed by more than 160 nations.

Severn Cullis-Suzuki speaks for six minutes to the delegates urging them to work hard on resolving global environmental issues. She received a standing ovation.

The Montreal Protocol is again amended with signatories agreeing to phase out CFC use by the year 2000.

1993 Sugar producers and U.S. government agree on a restoration plan for the Florida Everglades.

1994 *Dumping in Dixie: Class and Environmental Quality* is published by **Robert Bullard.** The book reports on five environmental justice campaigns in states ranging from Texas to West Virginia. Bullard emphasizes that African Americans are concerned about and do participate in environmental issues.

The California Desert Protection Act is passed.

Failure of a dike results in the release of 102,000 tons of oil into the Siberian tundra near Usink in northern Russia.

The Russian government calls for preventive measures to control the destruction of Lake Baikail.

The bald eagle is reclassified from an endangered species to a threatened species on the U.S. Endangered Species List.

An 8.5-million-gallon spill is discovered in Unocal's Guadalupe oil field in California.

1995 The U.S. Government reintroduces endangered wolves to Yellowstone Park.

1999 Scientists report that the human population of Earth now exceeds 6 billion people.

The peregrine falcon is removed from the U.S. Endangered Species List.

The *New Carissa* runs aground off the coast of Oregon, leaking some oil into Coos Bay. The tanker is later towed into the open ocean and sunk.

Beyond Globalization: Shaping a Sustainable Global Economy is published by Hazel Henderson.

Paul Hawken coauthors *Natural Capitalism, Creating the Next Industrial Revolution.*

Off the Map, an Expedition Deep into Imperialism, the Global Economy, and Other Earthly Whereabouts is published by **Chellis Glendinning**.

Twenty-three-year-old **Julia Butterfly Hill** comes down out of a 180-foot California redwood tree after living there for two years to prevent the destruction of the forest. A deal is made with the logging company to spare the tree as well as a three-acre buffer zone.

2000 Denis Hayes is the coordinator and **Mark Dubois** is the international coordinator of Earthday 2000.

Ralph Nader and **Winona LaDuke** run for U.S. president and vice president on the Green Party ticket.

In January 2000, Hazel Wolf passes away at the age of 101.

The Chernobyl nuclear power plant is scheduled to close in December.

Anthropologists for the Wildlife Conservation Society in New York announce that a type of large West African monkey is extinct, making it the first primate to vanish in the twenty-first century.

A study by National Park Trust, a privately funded land conservancy, finds that more than 90,000 acres within state parks in 32 states are threatened by commercial and residential development and increased traffic, among other things.

A bone-dry summer in north-central Texas breaks the Depression-era drought record when the Dallas area marks 59 days without rain. The arid streak with 100-degree daily highs breaks a record of 58 days set in the midst of the Dust Bowl in 1934 and tied in 1950. The Texas drought exceeded 1 billion dollars in agricultural losses.

Massachusetts announces that the state will spend $600,000 to determine whether petroleum pollution in largely African American city neighborhoods contributes to lupus, a potentially deadly immune disease. The research, to be conducted over three years, will target three areas of the city with unusually high levels of petroleum contamination.

Hybrid vehicle Toyota Prius is offered for sale in the United States.

The hole in the ozone layer over Antarctica has stretched over a populated city for the first time, after ballooning to a new record size. Previously, the hole had opened only over Antarctica and the surrounding ocean.

2001 An environmental group that successfully campaigned for the return of wolves to Yellowstone National Park wants the federal government to do the same in western Colorado and parts of Utah, southern Wyoming, northern New Mexico, and Arizona.

The UN Environment Program launches a campaign to save the world's great apes from extinction, asking for at least $1 million to get started.

The captain and crew of a tanker that spilled at least 185,000 gallons of diesel into the fragile marine environment of the Galapagos Islands have been arrested.

One hundred sixty-five countries approve the Kyoto rules aimed at halting global warming. The Kyoto Protocol requires industrial countries to scale back emissions of carbon dioxide and other greenhouse gases by an average of 5 percent from their 1990 levels by 2012. The United States, the world's biggest polluter rejects the pact.

The EPA reaches an agreement for the phaseout of a widely used pesticide, diazinon, because of potential health risks to children.

For the second time in three years, the average fuel economy of new passenger cars and light trucks sold in the United States dropped to its lowest level since 1980.

More and more Americans are breathing dirtier air, and larger U.S. cities such as Los Angeles and Atlanta remain among the worst for pollution.

In rural stretches of Alaska, global warming has thinned the Arctic pack ice, making travel dangerous for native hunters. Traces of industrial pollution from distant continents is showing up in the fat of Alaska's marine wildlife and in the breast milk of native mothers who eat a traditional diet including seal and walrus meat.

2002 A Congo volcano devastates a Congolese town burning everything in its path, creating a five-foot-high wall of cooling stone, and leaving a half million people homeless.

New research is conducted in the practice of killing sharks solely for their fins.

A report by the USGS shows the nation's waterways are awash in traces of chemicals used in beauty aids, medications, cleaners, and foods. Among the substances are caffeine, painkillers, insect repellent, perfumes, and nicotine. These substances largely escape regulation and defy municipal wastewater treatment.

A microbe is discovered to be a major cause of the destruction of beech trees in the northeastern United States.

A study discovers that, if fallen leaves are left in stagnant water, they can release toxic mercury, which eventually can accumulate in fish that live far downstream.

Scientists are experimenting with various sprays containing clay particles to kill toxic algae in seawater.

Meteorologists discover that the Mediterranean Sea receives air current pollutants from Europe, Asia, and North America.

Researchers report possible ways of blocking the deadly effects of anthrax.

2003 A new international treaty—The Protocol on Persistent Organic Pollutants (POPS) was ratified by 17 nations although the United States has not signed on. The treaty drafted by United Nations reduces and eliminates 16 toxic chemicals that are long-lived in the environment and travel globally. The new treaty, an extension of an earlier one signed in 2000, added four more organic persistent pollutants to the list.

Many global scientific studies reveal that excessive ultraviolet (UV) sunlight and pollution are linked to a decline in amphibian populations. Now Canadian biologists find that too much exposure of excessive UV radiation to tadpole populations reduces their chances of becoming frogs.

2003 marked the 50th anniversary of the research and publication of a different structure of the DNA model proposed by James D. Watson and

Francis H.C. Crick. In 1953 the scientists reported that the DNA molecule resembled a spiral staircase.

A new excavation in South Africa discovered the oldest fossils in the human family. The bones of a skull and a partial arm found in two caves date back to 4 million years ago according to scientists in Johannesburg

Scientists in New Jersey discovered that some outdoor antimosquito coils used to keep insects away can also cause respiratory health problems. The spiral-shaped container releases pollutants in the fumes expelled from coil. The researchers suggest that consumers should check these products carefully.

Researchers in Australia reported that pieces of plastic litter found in oceans continue to have an effect on marine wildlife. Small plastic chips are a hazard for seabirds who mistake the litter for food or fish eggs. The litter also moves up the food chain from fish that have ingested the plastic chips and in turn seals eat them.

2004 A scientific study reported that consumers should limit their consumption of farm-raised Atlantic salmon because of high concentrations of chlorinated organic contaminants in the fish. Their study revealed that the farm-raised salmon were contaminated with polychlorinated biphenyls (PCBs) and other organic chemicals. Except for the PCBs, the researchers agree that the farm-raised fish are healthy but consumption should be limited to no more than once a month in the diet. The researchers based their dietary report on the U.S. Environmental Protection Agency cancer risk assessments.

A group in Salisbury Plain, England is restoring Stonehenge to its natural setting. As a popular historic site to visitors, Stonehenge had become an area surrounded by roads and parking lots. The new restoration plan calls for building an underground tunnel for traffic and removing one of the roads. The present parking lots will become open grassy lawns.

Experts reported that two billion people lack reliable access to safe and nutritious food and 800 million, 40 percent of them children, are classified as chronically malnourished.

Public health officials in Uganda have reported progress in the country's fight against HIV, the AIDS virus. Since 1990's HIV cases in Uganda have dropped by more than 60 percent. Unfortunately, Uganda's neighboring countries are not doing well in their HIV prevention programs.

United Nations Secretary-General Kofi Annan stated, "by 2025, two-thirds of the world's population may be living in countries that face serious water shortages." The growing population is making surface water scarcer particularly in urban areas.

United States and Israel scientists have found a way to produce hydrogen from water. The hydrogen energy can be used in making fuel cells to power vehicles and homes. The research team uses solar radiation to heat sodium hydroxide in a solution of water. At high temperatures the water molecules (H_2O) break apart into oxygen and hydrogen. Using solar-power to produce hydrogen is better environmentally than hydrogen derived from fossil fuels.

Appendix B: Endangered Species by State

The list below, obtained from the U.S. Fish and Wildlife Service, is an abridged listing of a selected group of endangered species (E) for each state. For a full list of endangered and threatened species, and other information about endangered species and the Endangered Species Act, see the Endangered Species Program Website at http://endangered.fws.gov/

ALABAMA

(Alabama has 106 plant and animal species that are listed as endangered (E) or threatened (T). The following list is only a selection of those plants and animals that are endangered. Contact the U.S. Fish and Wildlife Service to see the entire list.)

Animals

E - Bat, gray
E - Bat, Indiana
E - Cavefish, Alabama
T - Chub, spotfin
E - Clubshell, black
E - Combshell, southern
E - Darter, boulder
E - Fanshell
E - Kidneyshell, triangular
E - Lampmussel, Alabama
E - Manatee, West Indian
E - Moccasinshell, Coosa
E - Mouse, Alabama beach
E - Mussel, ring pink
E - Pearlymussel, cracking
E - Pearlymussel, Cumberland monkeyface
E - Pigtoe, dark
E - Plover, piping
E - Shrimp, Alabama cave
E - Snail, tulotoma (Alabama live-bearing)
E - Stork, wood
E - Turtle, Alabama redbelly (red-bellied)
E - Turtle, leatherback sea
E - Woodpecker, red-cockaded

Plants

E - Grass, Tennessee yellow-eyed
E - Leather-flower, Alabama
E - Morefield's leather-flower
E - Pinkroot, gentian
E - Pitcher-plant, Alabama canebrake
E - Pitcher-plant, green
E - Pondberry
E - Prairie-clover, leafy

ALASKA

Animals

E - Curlew, Eskimo (*Numenius borealis*)
E - Falcon, American peregrine (*Falco peregrinus anatum*)

Plant

E - Aleutian shield-fern (Aleutian holly-fern) (*Polystichum aleuticum*)

ARIZONA

Animals

E - Ambersnail, Kanab
E - Bat, lesser (Sanborn's) long-nosed
E - Bobwhite, masked (quail)
E - Chub, bonytail
E - Chub, humpback
E - Chub, Virgin River
E - Chub, Yaqui
E - Flycatcher, Southwestern willow
E - Jaguarundi
E - Ocelot
E - Pronghorn, Sonoran
E - Pupfish, desert
E - Rail, Yuma clapper
E - Squawfish, Colorado
E - Squirrel, Mount Graham red
E - Sucker, razorback
E - Topminnow, Gila (incl. Yaqui)
E - Trout, Gila
E - Vole, Hualapai Mexican
E - Woundfin

Plants

E - Arizona agave
E - Arizona cliffrose
E - Arizona hedgehog cactus
E - Brady pincushion cactus
E - Kearney's blue-star
E - Nichol's Turk's head cactus
E - Peebles Navajo cactus
E - Pima pineapple cactus
E - Sentry milk-vetch

ARKANSAS

Animals

E - Bat, gray
E - Bat, Indiana
E - Bat, Ozark big-eared
E - Beetle, American burying (giant carrion)
E - Crayfish, cave
E - Pearlymussel, Curtis'
E - Pearlymussel, pink mucket
E - Pocketbook, fat
E - Pocketbook, speckled
E - Rock-pocketbook, Ouachita (Wheeler's pearly mussel)
E - Sturgeon, pallid
E - Tern, least
E - Woodpecker, red-cockaded

Plants

E - Harperella
E - Pondberry
E - Running buffalo clover

CALIFORNIA

(California has more than 160 plant and animal species that are listed as endangered or threatened. The following list is only a selection of those plants and animals that are endangered. Contact the U.S. Fish and Wildlife Service to see the entire list.)

Animals

E - Butterfly, El Segundo blue
E - Butterfly, Lange's metalmark
E - Chub, Mohave tui
E - Condor, California
E - Crayfish, Shasta (placid)
E - Fairy shrimp, Conservancy
E - Falcon, American peregrine
E - Fly, Delhi Sands flower-loving
E - Flycatcher, Southwestern willow
E - Fox, San Joaquin kit
E - Goby, tidewater
E - Kangaroo rat, Fresno
E - Lizard, blunt-nosed leopard
E - Mountain beaver, Point Arena
E - Mouse, Pacific pocket
E - Pelican, brown
E - Pupfish, Owens
E - Rail, California clapper
E - Salamander, Santa Cruz long-toed
E - Shrike, San Clemente loggerhead
E - Shrimp, California freshwater
E - Snail, Morro shoulderband (banded dune)
E - Snake, San Francisco garter
E - Stickleback, unarmored threespine
E - Sucker, Lost River
E - Tadpole shrimp, vernal pool
E - Tern, California least
E - Toad, arroyo southwestern
E - Turtle, leatherback sea
E - Vireo, least Bell's
E - Vole, Amargosa

Plants

E - Antioch Dunes evening-primrose
E - Bakersfield cactus
E - Ben Lomond wallflower
E - Burke's goldfields
E - California jewelflower
E - California Orcutt grass
E - Clover lupine
E - Cushenbury buckwheat
E - Fountain thistle
E - Gambel's watercress
E - Kern mallow
E - Loch Lomond coyote-thistle
E - Robust spineflower (includes Scotts Valley spineflower)
E - San Clemente Island larkspur
E - San Diego button-celery
E - San Mateo thornmint
E - Santa Ana River woolly-star
E - Santa Barbara Island liveforever
E - Santa Cruz cypress
E - Solano grass
E - Sonoma sunshine (Baker's stickyseed)

E - Stebbins' morning-glory
E - Truckee barberry
E - Western lily

COLORADO

Animals

E - Butterfly, Uncompahgre fritillary
E - Chub, bonytail
E - Chub, humpback
E - Crane, whooping
E - Ferret, black-footed
E - Flycatcher, Southwestern willow
E - Plover, piping
E - Squawfish, Colorado
E - Sucker, razorback
E - Tern, least
E - Wolf, gray

Plants

E - Clay-loving wild-buckwheat
E - Knowlton cactus
E - Mancos milk-vetch
E - North Park phacelia
E - Osterhout milk-vetch
E - Penland beardtongue

CONNECTICUT

Animals

E - Mussel, dwarf wedge
E - Plover, piping
E - Tern, roseate
E - Turtle, hawksbill sea
E - Turtle, Kemp's (Atlantic) ridley sea
E - Turtle, leatherback sea

Plant

E - Sandplain gerardia

DELAWARE

Animals

E - Plover, piping
E - Squirrel, Delmarva Peninsula fox
E - Turtle, hawksbill sea
E - Turtle, Kemp's (Atlantic) ridley sea—Turtle, green sea
E - Turtle, leatherback sea

Plant

E - Canby's dropwort

FLORIDA

(Florida has more than 90 plant and animal species that are listed as endangered or threatened. The following list is only a selection of those plants and animals that are endangered. Contact the U.S. Fish and Wildlife Service to see the entire list.)

Animals

E - Bat, gray
E - Butterfly, Schaus swallowtail
E - Crocodile, American
E - Darter, Okaloosa
E - Deer, key
E - Kite, Everglade snail
E - Manatee, West Indian (Florida)
E - Mouse, Anastasia Island beach
E - Mouse, Choctawahatchee beach
E - Panther, Florida
E - Plover, piping
E - Rabbit, Lower Keys
E - Rice rat (silver rice rat)
E - Sparrow, Cape Sable seaside
E - Stork, wood
E - Tern, roseate
E - Turtle, hawksbill sea
E - Turtle, Kemp's (Atlantic) ridley sea
E - Turtle, leatherback sea
E - Vole, Florida salt marsh
E - Woodpecker, red-cockaded
E - Woodrat, Key Largo

Plants

E - Apalachicola rosemary
E - Beautiful pawpaw
E - Brooksville (Robins') bellflower
E - Carter's mustard
E - Chapman rhododendron
E - Cooley's water-willow
E - Crenulate lead-plant
E - Etonia rosemary
E - Florida golden aster
E - Fragrant prickly-apple
E - Garrett's mint
E - Key tree-cactus
E - Lakela's mint
E - Okeechobee gourd

E - Scrub blazingstar
E - Small's milkpea
E - Snakeroot
E - Wireweed

GEORGIA

Animals

E - Acornshell, southern
E - Bat, gray
E - Bat, Indiana
E - Clubshell, ovate
E - Clubshell, southern
E - Combshell, upland
E - Darter, amber
E - Darter, Etowah
E - Kidneyshell, triangular
E - Logperch, Conasauga
E - Manatee, West Indian (Florida)
E - Moccasinshell, Coosa
E - Pigtoe, southern
E - Plover, piping
E - Stork, wood
E - Turtle, hawksbill sea
E - Turtle, Kemp's (Atlantic) ridley sea
E - Turtle, leatherback sea
E - Woodpecker, red-cockaded

Plants

E - American chaffseed
E - Black-spored quillwort
E - Canby's dropwort
E - Florida torreya
E - Fringed campion
E - Green pitcher-plant
E - Hairy rattleweed
E - Harperella
E - Large-flowered skullcap
E - Mat-forming quillwort
E - Michaux's sumac
E - Persistent trillium
E - Pondberry
E - Relict trillium
E - Smooth coneflower
E - Tennessee yellow-eyed grass

HAWAII

(Hawaii has 300 plant and animal species listed as endangered or threatened. The following list is only a selection of those plants and animals that are endangered. Contact the U.S. Fish and Wildlife Service to see the entire list.)

Animals

E - 'Akepa, Hawaii (honeycreeper)
E - Bat, Hawaiian hoary
E - Coot, Hawaiian
E - Creeper, Hawaiian
E - Crow, Hawaiian
E - Duck, Hawaiian
E - Duck, Laysan
E - Finch, Laysan (honeycreeper)
E - Finch, Nihoa (honeycreeper)
E - Goose, Hawaiian (nene)
E - Hawk, Hawaiian
E - Millerbird, Nihoa (old world warbler)
E - Nukupu'u (honeycreeper)
E - Palila (honeycreeper)
E - Parrotbill, Maui (honeycreeper)
E - Petrel, Hawaiian dark-rumped
E - Snails, Oahu tree
E - Stilt, Hawaiian
E - Turtle, hawksbill sea
E - Turtle, leatherback sea

Plants

E - Abutilon eremitopetalum
E - Bonamia menziesii
E - Carter's panicgrass
E - Diamond Head schiedea
E - Dwarf iliau
E - Fosberg's love grass
E - Hawaiian bluegrass
E - Hawaiian red-flowered geranium
E - Kaulu
E - Kiponapona
E - Mahoe
E - Mapele
E - Nanu
E - Nehe
E - Opuhe
E - Pamakani
E - Round-leaved chaff-flower
E - Viola helenae

IDAHO

Animals

E - Caribou, woodland
E - Crane, whooping

E - Limpet, Banbury Springs
E - Snail, Snake River physa
E - Snail, Utah valvata
E - Springsnail, Bruneau Hot
E - Springsnail, Idaho
E - Sturgeon, white
E - Wolf, gray

Plants

(No plants on the endangered list)

ILLINOIS

Animals

E - Bat, gray
E - Bat, Indiana
E - Butterfly, Karner blue
E - Dragonfly, Hine's emerald
E - Falcon, American peregrine
E - Fanshell
E - Pearlymussel, Higgins' eye
E - Pearlymussel, orange-foot pimple back
E - Pearlymussel, pink mucket
E, T - Plover, piping
E - Pocketbook, fat
E - Snail, Iowa Pleistocene
E - Sturgeon, pallid
E - Tern, least

Plant

E - Leafy prairie-clover

INDIANA

Animals

E - Bat, gray
E - Bat, Indiana
E - Butterfly, Karner blue
E - Butterfly, Mitchell's satyr
E - Clubshell
E - Fanshell
E - Mussel, ring pink (golf stick pearly)
E - Pearlymussel, cracking
E - Pearlymussel, orange-foot pimple back
E - Pearlymussel, pink mucket
E - Pearlymussel, tubercled-blossom
E - Pearlymussel, white cat's paw
E - Pearlymussel, white wartyback
E - Pigtoe, rough
E, T - Plover, piping
E - Pocketbook, fat
E - Riffleshell, northern
E - Tern, least

Plant

E - Running buffalo clover

IOWA

Animals

E - Bat, Indiana
E - Pearlymussel, Higgins' eye
E - Plover, piping
E - Snail, Iowa Pleistocene
E - Sturgeon, pallid
E - Tern, least

Plants

(No plants on the endangered list)

KANSAS

Animals

E - Bat, gray
E - Bat, Indiana
E - Crane, whooping
E - Curlew, Eskimo
E - Ferret, black-footed
E - Plover, piping
E - Sturgeon, pallid
E - Tern, least
E - Vireo, black-capped

Plants

(No plants on endangered list)

KENTUCKY

Animals

E - Bat, gray
E - Bat, Indiana
E - Bat, Virginia big-eared
E - Clubshell
E - Darter, relict
E - Falcon, American peregrine
E - Fanshell
E - Mussel, ring pink (golf stick pearly)
E - Mussel, winged mapleleaf
E - Pearlymussel, cracking
E - Pearlymussel, Cumberland bean

E - Pearlymussel, dromedary
E - Pearlymussel, little-wing
E - Pearlymussel, orange-foot pimple back
E - Pearlymussel, pink mucket
E - Pearlymussel, purple cat's paw
E - Pearlymussel, tubercled-blossom
E - Pearlymussel, white wartyback
E - Pigtoe, rough
E - Plover, piping
E - Pocketbook, fat
E - Riffleshell, northern
E - Riffleshell, tan
E - Shiner, Palezone
E - Shrimp, Kentucky cave
E - Sturgeon, pallid
E - Tern, least
E - Woodpecker, red-cockaded

Plants

E - Cumberland sandwort
E - Rock cress
E - Running buffalo clover
E - Short's goldenrod

LOUISIANA

Animals

E - Manatee, West Indian (Florida)
E - Pearlymussel, pink mucket
E - Pelican, brown
E - Plover, piping
E - Sturgeon, pallid
E - Tern, least
T - Turtle, green sea
E - Turtle, hawksbill sea
E - Turtle, Kemp's (Atlantic) ridley sea
E - Turtle, leatherback sea
E - Vireo, black-capped
E - Woodpecker, red-cockaded

Plants

E - American chaffseed
E - Louisiana quillwort
E - Pondberry

MAINE

Animals

E - Plover, piping
E - Tern, roseate
E - Turtle, leatherback sea

Plant

E - Furbish lousewort

MARYLAND

Animals

E - Bat, Indiana
E - Darter, Maryland
E - Mussel, dwarf wedge
E - Plover, piping
E - Squirrel, Delmarva Peninsula fox
E - Turtle, hawksbill sea
E - Turtle, Kemp's (Atlantic) ridley sea
E - Turtle, leatherback sea

Plants

E - Canby's dropwort
E - Harperella
E - Northeastern (Barbed bristle) bulrush
E - Sandplain gerardia

MASSACHUSETTS

Animals

E - Beetle, American burying (giant carrion)
E - Falcon, American peregrine
E - Mussel, dwarf wedge
E - Plover, piping
E - Tern, roseate
E - Turtle, hawksbill sea
E - Turtle, Kemp's (Atlantic) ridley sea
E - Turtle, leatherback sea
E - Turtle, Plymouth redbelly (red-bellied)

Plants

E - Northeastern (Barbed bristle)
E - Sandplain gerardia

MICHIGAN

Animals

E - Bat, Indiana
E - Beetle, American burying (giant carrion)
E - Beetle, Hungerford's crawling water
E - Butterfly, Karner blue
E - Butterfly, Mitchell's satyr
E - Clubshell
E - Plover, piping
E - Riffleshell, northern
E - Warbler, Kirtland's
E - Wolf, gray

Plant

E - Michigan monkey-flower

MINNESOTA

Animals

E - Butterfly, Karner blue
E - Mussel, winged mapleleaf
E - Pearlymussel, Higgins' eye
E - Plover, piping
E - Wolf, gray

Plant

E - Minnesota trout lily

MISSISSIPPI

Animals

E - Bat, Indiana
E - Clubshell, black (Curtus' mussel)
E - Clubshell, ovate
E - Clubshell, southern
E - Combshell, southern (penitent mussel)
E - Crane, Mississippi sandhill
E - Falcon, American peregrine
E - Manatee, West Indian (Florida)
E - Pelican, brown
E - Pigtoe, flat (Marshall's mussel)
E - Pigtoe, heavy (Judge Tait's mussel)
E - Plover, piping
E - Pocketbook, fat
E - Stirrupshell
E - Sturgeon, pallid
E - Tern, least
E - Turtle, hawksbill sea
E - Turtle, Kemp's (Atlantic) ridley sea
E - Turtle, leatherback sea
E - Woodpecker, red-cockaded

Plants

E - American chaffseed
E - Pondberry

MISSOURI

Animals

E - Bat, gray
E - Bat, Indiana
E - Bat, Ozark big-eared
E - Pearlymussel, Curtis'
E - Pearlymussel, Higgins' eye
E - Pearlymussel, pink mucket
E - Plover, piping
E - Pocketbook, fat
E - Sturgeon, pallid
E - Tern, least

Plants

E - Missouri bladderpod
E - Pondberry
E - Running buffalo clover

MONTANA

Animals

E - Crane, whooping
E - Curlew, Eskimo
E - Ferret, black-footed
E - Plover, piping
E - Sturgeon, pallid
E - Sturgeon, white
E - Tern, least
E - Wolf, gray

Plants

(No plants on endangered list)

NEBRASKA

Animals

E - Beetle, American burying (giant carrion)
E - Crane, whooping
E - Curlew, Eskimo
E - Ferret, black-footed
E - Plover, piping
E - Sturgeon, pallid
E - Tern, least

Plant

E - Blowout penstemon

NEVADA

Animals

E - Chub, bonytail
E - Chub, Pahranagat roundtail (bonytail)
E - Chub, Virgin River
E - Cui-ui
E - Dace, Ash Meadows speckled

E - Dace, Clover Valley speckled
E - Dace, Independence Valley speckled
E - Dace, Moapa
E - Poolfish (killifish), Pahrump
E - Pupfish, Ash Meadows Amargosa
E - Pupfish, Devils Hole
E - Pupfish, Warm Springs
E - Spinedace, White River
E - Springfish, Hiko White River
E - Springfish, White River
E - Sucker, razorback
E - Woundfin

Plants

E - Amargosa niterwort
E - Steamboat buckwheat

NEW HAMPSHIRE

Animals

E - Butterfly, Karner blue
E - Mussel, dwarf wedge
E - Turtle, leatherback sea

Plants

E - Jesup's milk-vetch
E - Northeastern (Barbed bristle) bulrush
E - Robbins' cinquefoil

NEW JERSEY

Animals

E - Bat, Indiana
E - Plover, piping
E - Tern, roseate
E - Turtle, hawksbill sea
E - Turtle, Kemp's (Atlantic) ridley sea
E - Turtle, leatherback sea

Plant

E - American chaffseed

NEW MEXICO

Animals

E - Bat, lesser (Sanborn's) long-nosed
E - Bat, Mexican long-nosed
E - Crane, whooping
E - Gambusia, Pecos
E - Isopod, Socorro
E - Minnow, Rio Grande silvery
E - Springsnail, Alamosa
E - Springsnail, Socorro
E - Sucker, razorback
E - Tern, least
E - Topminnow, Gila (incl. Yaqui)
E - Trout, Gila
E - Woundfin

Plants

E - Holy Ghost ipomopsis
E - Knowlton cactus
E - Kuenzler hedgehog cactus
E - Lloyd's hedgehog cactus
E - Mancos milk-vetch
E - Sacramento prickly-poppy
E - Sneed pincushion cactus
E - Todsen's pennyroyal

NEW YORK

Animals

E - Butterfly, Karner blue
E - Mussel, dwarf wedge
E, T - Plover, piping
E - Tern, roseate
E - Turtle, hawksbill sea
E - Turtle, Kemp's (Atlantic) ridley sea
E - Turtle, leatherback sea

Plants

E - Northeastern (Barbed bristle) bulrush
E - Sandplain gerardia

NORTH CAROLINA

Animals

E - Bat, Indiana
E - Bat, Virginia big-eared
E - Butterfly, Saint Francis' satyr
E - Elktoe, Appalachian
E - Falcon, American peregrine
E - Heelsplitter, Carolina
E - Manatee, West Indian (Florida)
E - Mussel, dwarf wedge
E - Pearlymussel, little-wing
E - Plover, piping
E - Shiner, Cape Fear
E - Spider, spruce-fir moss

E - Spinymussel, Tar River
E - Squirrel, Carolina northern flying
E - Tern, roseate
E - Turtle, hawksbill sea
E - Turtle, Kemp's (Atlantic) ridley sea
E - Turtle, leatherback sea
E - Wolf, red
E - Woodpecker, red-cockaded

Plants

E - American chaffseed
E - Bunched arrowhead
E - Canby's dropwort
E - Cooley's meadowrue
E - Green pitcher-plant
E - Harperella
E - Michaux's sumac
E - Mountain sweet pitcher-plant
E - Pondberry
E - Roan Mountain bluet
E - Rock gnome lichen
E - Rough-leaved loosestrife
E - Schweinitz's sunflower
E - Small-anthered bittercress
E - Smooth coneflower
E - Spreading avens
E - White irisette

NORTH DAKOTA

Animals

E - Crane, whooping
E - Curlew, Eskimo
E - Falcon, American peregrine
E - Ferret, black-footed
E - Plover, piping
E - Sturgeon, pallid
E - Tern, least
E - Wolf, gray

Plants

(No plants on endangered list)

OHIO

Animals

E - Bat, Indiana
E - Beetle, American burying (giant carrion)
E - Butterfly, Karner blue
E - Butterfly, Mitchell's satyr
E - Clubshell
E - Dragonfly, Hine's emerald
E - Fanshell
E - Madtom, Scioto
E - Pearlymussel, pink mucket
E - Pearlymussel, purple cat's paw
E - Pearlymussel, white cat's paw
E,T - Plover, piping
E - Riffleshell, northern

Plant

E - Running buffalo clover

OKLAHOMA

Animals

E - Bat, gray
E - Bat, Indiana
E - Bat, Ozark big-eared
E - Beetle, American burying (giant carrion)
E - Crane, whooping
E - Curlew, Eskimo
E - Plover, piping
E - Rock-pocketbook, Ouachita
E - Tern, least
E - Vireo, black-capped
E - Woodpecker, red-cockaded

Plants

(No plants on the endangered list)

OREGON

Animals

E - Chub, Borax Lake
E - Chub, Oregon
E - Deer, Columbian white-tailed
E - Pelican, brown
E - Sucker, Lost River
E - Sucker, shortnose
E - Turtle, leatherback sea

Plants

E - Applegate's milk-vetch
E - Bradshaw's desert-parsley
E - Malheur wire-lettuce
E - Marsh sandwort
E - Western lily

PENNSYLVANIA

Animals

E - Bat, Indiana
E - Clubshell
E - Mussel, dwarf wedge
E - Mussel, ring pink (golf stick pearly)
E - Pearlymussel, cracking
E - Pearlymussel, orange-foot pimple back
E - Pearlymussel, pink mucket
E - Pigtoe, rough
E,T - Plover, piping
E - Riffleshell, northern

Plant

E - Northeastern (Barbed bristle) bulrush

RHODE ISLAND

Animals

E - Beetle, American burying
E - Falcon, American peregrine
E - Plover, piping
E - Tern, roseate
E - Turtle, hawksbill sea
E - Turtle, Kemp's
E - Turtle, leatherback sea

Plant

E - Sandplain gerardia

SOUTH CAROLINA

Animals

E - Bat, Indiana
E - Heelsplitter, Carolina
E - Manatee, West Indian (Florida)
E - Plover, piping
E - Stork, wood
E - Tern, roseate
E - Turtle, hawksbill sea
E - Turtle, Kemp's (Atlantic) ridley sea
E - Turtle, leatherback sea
E - Woodpecker, red-cockaded

Plants

E - American chaffseed
E - Black-spored quillwort
E - Bunched arrowhead
E - Canby's dropwort
T - Dwarf-flowered heartleaf
E - Harperella
E - Michaux's sumac
E - Mountain sweet pitcher-plant
E - Persistent trillium
E - Pondberry
E - Relict trillium
E - Rough-leaved loosestrife
E - Schweinitz's sunflower
E - Smooth coneflower

SOUTH DAKOTA

Animals

E - Beetle, American burying (giant carrion)
E - Crane, whooping
E - Curlew, Eskimo
E - Ferret, black-footed
E - Plover, piping
E - Sturgeon, pallid
E - Tern, least
E - Wolf, gray

Plants

(No plants on the endangered list)

TENNESSEE

(Tennessee has 81 plant and animal species that are listed as endangered or threatened. The following list is only a selection of those plants and animals that are endangered. Contact the U.S. Fish and Wildlife Service to see the entire list.)

Animals

E - Bat, gray
E - Bat, Indiana
E - Combshell, upland
E - Crayfish, Nashville
E - Darter, amber
E - Fanshell
E - Lampmussel, Alabama
E - Madtom, Smoky
E - Marstonia (snail), (royalobese)
E - Moccasinshell, Coosa
E - Mussel, ring pink (golf stick pearly)
E - Pearlymussel, Appalachian monkeyface

E - Pearlymussel, Cumberland bean
E - Riversnail, Anthony's
E - Spider, spruce-fir moss
E - Squirrel, Carolina northern flying
E - Sturgeon, pallid
E - Tern, least
E - Wolf, red
E - Woodpecker, red-cockaded

Plants

E - Cumberland sandwort
E - Green pitcher-plant
E - Large-flowered skullcap
E - Leafy prairie-clover (Dalea)
E - Roan Mountain bluet
E - Rock cress
E - Rock gnome lichen
E - Ruth's golden aster
E - Spring Creek bladderpod
E - Tennessee purple coneflower
E - Tennessee yellow-eyed grass

TEXAS

(Texas has 70 plant and animal species that are listed as endangered or threatened. The following list is only a selection of those plants and animals that are endangered. Contact the U.S. Fish and Wildlife Service to see the entire list.)

Animals

E - Bat, Mexican long-nosed
E - Beetle, Coffin Cave mold
E - Crane, whooping
E - Curlew, Eskimo
E - Darter, fountain
E - Falcon, northern aplomado
E - Jaguarundi
E - Manatee, West Indian (Florida)
E - Minnow, Rio Grande silvery
E - Ocelot
E - Pelican, brown
E - Plover, piping
E - Prairie-chicken, Attwater's greater
E - Pupfish, Comanche Springs
E - Salamander, Texas blind
E - Spider, Tooth Cave
E - Tern, least
E - Toad, Houston
E - Turtle, hawksbill sea
E - Turtle, Kemp's (Atlantic) ridley sea
E - Vireo, black-capped
E - Warbler, golden-cheeked
E - Woodpecker, red-cockaded

Plants

E - Ashy dogweed
E - Black lace cactus
T - Hinckley's oak
E - Large-fruited sand-verbena
E - Little Aguja pondweed
E - Lloyd's hedgehog cactus
E - Nellie cory cactus
E - Sneed pincushion cactus
E - South Texas ambrosia
E - Star cactus
E - Terlingua Creek cats-eye
E - Texas poppy-mallow
E - Texas snowbells
E - Texas wild-rice
E - Tobusch fishhook cactus
E - Walker's manioc

UTAH

Animals

E - Ambersnail, Kanab
E - Chub, bonytail
E - Chub, humpback
E - Chub, Virgin River
E - Crane, whooping
E - Ferret, black-footed
E - Flycatcher, Southwestern willow
E - Snail, Utah valvata
E - Squawfish, Colorado
E - Sucker, June
E - Sucker, razorback
E - Woundfin

Plants

E - Autumn buttercup
E - Barneby reed-mustard
E - Barneby ridge-cress (peppercress)
E - Clay phacelia
E - Dwarf bear-poppy
E - Kodachrome bladderpod
E - San Rafael cactus
E - Shrubby reed-mustard (toad-flax cress)
E - Wright fishhook cactus

VERMONT

Animals

E - Bat, Indiana
E - Mussel, dwarf wedge

Plants

E - Jesup's milk-vetch
E - Northeastern (Barbed bristle) bulrush

VIRGINIA

Animals

E - Bat, gray
E - Bat, Indiana
E - Bat, Virginia big-eared
E - Darter, duskytail
E - Falcon, American peregrine
E - Fanshell
E - Isopod, Lee County cave
E - Logperch, Roanoke
E - Mussel, dwarf wedge
E - Pearlymussel, Appalachian monkeyface
E - Pearlymussel, birdwing
E - Pearlymussel, cracking
E - Pearlymussel, Cumberland monkeyface
E - Pearlymussel, dromedary
E - Pearlymussel, green-blossom
E - Pearlymussel, little-wing
E - Pearlymussel, pink mucket
E - Pigtoe, fine-rayed
E - Pigtoe, rough
E - Pigtoe, shiny
E - Plover, piping
E - Riffleshell, tan
E - Salamander, Shenandoah
E - Snail, Virginia fringed mountain
E - Spinymussel, James River (Virginia)
E - Squirrel, Delmarva Peninsula fox
E - Squirrel, Virginia northern flying
E - Turtle, hawksbill sea
E - Turtle, Kemp's (Atlantic) ridley sea
E - Turtle, leatherback sea
E - Woodpecker, red-cockaded

Plants

E - Northeastern (Barbed bristle) bulrush
E - Peter's Mountain mallow
E - Shale barren rock-cress
E - Smooth coneflower

WASHINGTON

Animals

E - Caribou, woodland
E - Deer, Columbian white-tailed
E - Pelican, brown
E - Turtle, leatherback sea
E - Wolf, gray

Plants

E - Bradshaw's desert-parsley (lomatium)
E - Marsh sandwort

WEST VIRGINIA

Animals

E - Bat, Indiana
E - Bat, Virginia big-eared
E - Clubshell
E - Falcon, American peregrine
E - Fanshell
E - Mussel, ring pink
E - Pearlymussel, pink mucket
E - Pearlymussel, tubercled-blossom
E - Riffleshell, northern
E - Spinymussel, James River
E - Squirrel, Virginia northern flying

Plants

E - Harperella
E - Northeastern (Barbed bristle) bulrush
E - Running buffalo clover
E - Shale barren rock-cress

WISCONSIN

Animals

E - Butterfly, Karner blue
E - Dragonfly, Hine's emerald
E - Mussel, winged mapleleaf
E - Pearlymussel, Higgins' eye
E, T - Plover, piping
E - Warbler, Kirtland's
E - Wolf, gray

Plants

(No plants on endangered list)

WYOMING

Animals

E - Crane, whooping
E - Dace, Kendall Warm Springs
E - Ferret, black-footed
E - Squawfish, Colorado
E - Sucker, razorback
E - Toad, Wyoming
E - Wolf, gray

Plants

(No plants on endangered list)

APPENDIX C: WEBSITES BY CLASSIFICATION

Please note that the authors have made a consistent effort to include up-to-date Websites. However, over time, some Websites may move or no longer be posted.

ACID MINE DRAINAGE

National Reclamation Center, West Virginia University, Evansdale office, http//www.nrcce.wvu.edu/.

ACID RAIN

http://www.epa.gov/docs/acidrain/andhome/html.

The EPA has a hotline to request educational materials or respond to questions regarding acid rain: (202) 343–9620. http://www.econet.apc.org/acid rain.

Environmental Protection Agency, http://www.epa.gov/docs/acidrain/effects/enveffct.html.

National Reclamation Center's West Virginia University, Evansdale office: http://www.nrcce.wvu.edu/

USGS Water Science/Acid Rain, http://wwwga.usgs.gov/edu/acidrain.html.

AGENCY FOR TOXIC SUBSTANCES AND DISEASES

Registry Division of Toxicology
1600 Clifton Road NE Mailstop E-29
Atlanta, GA 30333
Website: http://www.atsdr1.atsdr.cdc.gov:8080/atsdrhome.html.

Agency for Toxic Substances and Disease Registry, http://www.atsdr.cdc.gov/cxcx3.html.

Information on biosphere reserves and UNESCO's Man and the Biosphere Programme, UNESCO: http://www.unesco. org

Man and the Biosphere Program: http://www. mabnet.org

AGRICULTURE

United States Department of Agriculture, http://www.usda.gov.

ALTERNATIVE FUELS

Department of Energy, http://www.doe.gov.

Department of Energy Alternative Fuels Data Center, http://www.afdc.nrel.gov; http://www.afdc.doe.gov/; or http://www.fleets.doe.gov.

AMPHIBIANS

http://www.frogweb.gov/

ANTARCTICA

Antarctica Treaty, http://www.sedac.ciesin.org/pidb/register/reg-024.rrr.html.

Greenpeace International Antarctic Homepage, http://www.greenpeace.org/~comms/98/antarctic.

International Centre for Antarctic Information and Research Homepage (includes text of Antarctic Treaty), http://www.icair.iac.org.nz.

Virtual Antarctica, http://www.exploratorium.edu

ARCTIC

Arctic Circle (University of Connecticut), http://arcticcircle.uconn.edu/arcticcircle.

Arctic Council Home Page, http://www.nrc.ca/arctic/index.html.

Arctic Monitoring and Assessment Programme (Norway), http://www.gsf.de/ UNEP/amap1.html.

Arctic National Wildlife Refuge, http://energy.usgs.gov/factsheets/ANWR/ANWR.html.

Institute of Arctic and Alpine Research, http://instaar.colorado.edu.

Institute of the North (Alaska Pacific University),

Inuit Circumpolar Conference,
NOAA Fisheries, http://www.nmfs.gov/.

Nunavut,
Smithsonian Institution Arctic Studies Center, http://www.mnh.si.edu/arctic.

U.S. Fish and Wildlife Service
U.S. Department of the Interior

1849 C Street, NW,
Washington, D.C. 20240
Telephone: (202) 208-5634
Website: http://www.fws.gov.

World Conservation Monitoring Centre Arctic Programme, http://www.wcmc.org.uk/arctic.

AUTOMOBILE

Cars and Their Enviromental Impact, http://www.environment.volvocars.com/ch1-1.htm.

National Center for Vehicle Emissions Control and Safety (NCVECS), http://www.colostate.edu/Depts/NCVECS/ncvecs1.html.

U.S. Environmental Protection Agency Fact Sheet (EPA 400-F-92-004, August 1994), "Air Toxics from Motor Vehicles," http://www.epa.gov/oms/02-toxic.htm.

U.S. Enviromental Protection Agency, Office of Mobile Sources, http://www.epa.gov/oms.

BIOLOGICAL WEAPONS

Federation of American Scientist Biological Weapons Control, http://www.fas.org/bwc.

Chemical and Biological Defense Information Analysis Center, http://www.cbiac.apgea.army. mil

BIOMES

Committee for the National Institute for the Environment, http://www.cnie.org/nle/biodv-6.html.

BIOREMEDIATION

Consortium, http://www.rtdf.org/public/biorem.

BROWNFIELD

Projects, http://www.epa.gov/brownfields/.

CERES

Website: http://www.ceres.org or
e-mail ceres@igc.apc.org.

Summaries of Major Environmental Laws, http://www.epa.gov/region5/defs/index.html.

CHEETAHS

Cheetah Conservation Fund

4649 Sunnyside Avenue N, Suite 325
Seattle, WA 98103
Website: http://www.cheetah.org.

World Wildlife Fund

1250 24th Street, NW,
Washington, D.C. 20037
Telephone: 1-800-225-5993
Website: http://www.worldwildlife.org/.

CHEMICAL WEAPONS

Chemical Stockpile Disposal Project (CSDP), http://www.pmcd.apgea.army.mil/graphical/CSDP/index.html.

Tooele Chemical Agent Disposal Site Facility, http://www.deq.state.ut.us/eqshw/cds/tocdfhp1.htm.

CLEAN WATER ACT

Sierra Club, "Happy 25th Birthday, Clean Water Act," http://sierraclub.org/wetlands/cwabday.html.

CLIMATE CHANGE AND GLOBAL WARMING

U.S. Geological Survey, Climate Change and History, http://geology.usgs.gov/index.shtml.

EPA Global Warming Site, http://www.epa.gov/globalwarming.

Greenpeace International, Climate, http://www.greenpeace.org/~climate.

United Nations Intergovernmental Panel on Climate Change, http://www.ipcc.ch.

COAL

Coal Age Magazine, http://coalage.com.

Department of Energy, Office of Fossil Energy, http:/www.doe.gov.

U.S. Geological Survey, National Coal Resources Data System, http:energy.er.usgs.gov/coalqual. htm.

COASTAL AND MARINE GEOLOGY

U.S. Geological Survey, http://marine.usgs.gov/.

COMPOSTING

EPA Office of Solid Waste and Emergency Response—Composting, http:www.epa.gov/epaoswer/non-hw/compost/index.htm

Cornell Composting, http://www.cfe.cornell.edu/compost/Composting_Homepage.html

CONSENT DECREES

EPA Office of Enforcement and Compliance Assurance, http://es.epa.gov/oeca/osre/decree.html.

CORAL REEFS

Coral Reef Alliance, http://www.coral.org.

Coral Reef Network Directory, Greenpeace
1436 U Street, NW
Washington, D.C. 20009
Website: http://www.greenpeace.org.

EARTHDAY 2000

Earth Day Network

91 Marion Street,
Seattle, WA 98104
Telephone: 1(206)-264-0114.
Website: http://www.earthday.net/; and worldwide@earthday.net.

EARTHWATCH

Earthwatch Institute International, http://www. earthwatch.org.

EL NIÑO

El Niño/La Niña theme page, contact NOAA Website: http://www.pmel.noaa.gov/toga-tao/el-nino/nino-home-low.html.

NOAA, La Niña homepage, www.elnino.noaa.gov/lanina.html.

National Center for Atmospheric Research, http://www.ncar.ucar.edu/.

National Hurricane Center/Tropical Prediction Center, http://www.nhc.noaa.gov/.

National Oceanographic and Atmospheric Administration, http://www.noaa.gov/.

Scripps Institute of Oceanography, http://sio.ucsd.edu/supp_groups/siocomm/elnino/elnino.html.

ELECTRIC VEHICLES

Electric Vehicle Association of the Americas 800-438-3228, http://www.evaa.org.

Electric Vehicle Technology, http://www.avere.org/.

ELEPHANTS

African Wildlife Foundation, http://www.awf.org.

U.S. Fish and Wildlife Service, Species List of Endangered and Threatened Wildlife, http://endangered.fws.gov/

World Wildlife Fund, http://www.wwf.org.

ETHANOL

U.S. Department of Energy, Energy Efficiency and Renewable Energy Clearinghouse,

P.O. Box 3048
Merrifield, VA 22116
E-mail: energyinfo@delphi.com.
Website: http://www.doe.gov.

EVERGLADES

National Park Service, Everglades National Park, http://www.nps.gov/ever.

FEDERAL EMERGENCY MANAGEMENT AGENCY (FEMA)

FEMA, http://www.fema.gov.

FISHING, COMMERCIAL

National Oceanographic and Atmospheric Administration Fisheries, http://www.nmfs. gov/.

United Nations Food and Agriculture Organization Fisheries, http://www.fao.org/waicent/faoinfo/fishery/fishery.htm.

FORESTS

American Forests, http://www.amfor.org.

Greenpeace International, Forests, http://www. greenpeace.org/~forests.

Society of American Foresters, http://www. safnet.org.

U.S. Forest Service, http://www.fs.fed.us.

U.S. Forest Service Research, http://www.fs.fed.us/links/research.shtml.

World Conservation Monitoring Centre, http://www.wcmc.org.uk.

World Resources Institute Forest Frontiers Initiative, http://www.wri.org/ffi.

World Wildlife Fund (Worldwide Fund for Nature) Forests for Life Campaign, http://www.panda.org/forests4life.

FUEL CELLS AND OTHER ALTERNATIVE FUELS

Crest's Guide to the Internet's Alternative Energy Resources, http://solstice.crest.org/online/aeguide/aehome.html.

U.S. Department of Energy

P.O. Box 12316
Arlington, VA 22209
Telephone: 1-800-423-1363
Website: http://www.doe.gov.

U.S. Department of Energy, Alternative Fuels Data Center, http://www.afdc.nrel.gov.

GEOLOGY

Geological surveys, U.S. Geological Survey, http://www.usgs.gov/.

For general interest publications and products, http://mapping.usgs.gov/www/products/mappubs.html.

GEOTHERMAL SITES

Energy and Geoscience Institute

University of Utah
423 Wakara Way
Salt Lake City, UT 84108
Website: http://www.egi.utah.edu.

Geothermal energy information, http://geothermal.marin.org.

Geothermal database USA and Worldwide, http://www.geothermal.org.

International geothermal, http://www.demon.co.uk/geosci/igahome.html.

Solstice is the Internet information service of the Center for Renewable Energy and Sustainable Technology (CREST), http://solstice.crest.org/

GLACIERS SHRINKING

United States Geological Survey, Climate Change and History, http://geology.usgs.gov/index. shtml.

Sierra Club, Public Information Center, (415) 923-5653; or the Global Warming and Energy Team, (202) 547-1141, or by E-mail: information@sierraclub.org.

GLOBEC

Educational Website, http://cbl.umces.edu/fogarty/usglobec/misc/education.html.

GRASSLANDS AND PRAIRIES

Postcards from the Prairie, http://www.nrwrc.usgs.gov/postcards/postcards.htm.

University of California, Berkeley, World Biomes, Grasslands, http://www.ucmp.berkeley.edu/glossary/gloss5/biome/grasslan.html.

Worldwide Fund for Nature, Grasslands and Its Animals, http://www.panda.org/kids/wildlife/idxgrsmn.htm.

GROUNDWATER

EPA, http://www.epa.gov/swerosps/ej/.

Groundwater atlas of the United States, http://www.capp.er.usgs.gov/publicdocs/gwa/.

HAZARDOUS MATERIALS TRANSPORTAION ACT

Website: http://www.dot.gov.

HAZARDOUS SUBSTANCES

U.S. Environmental Protection Agency Program, http://epa.gov/.

U.S. Occupational Safety and Health Administration (OSHA), http://www.osha.gov/toxicsubstances/index.html.

Environmental Defense Fund (data on wastes and chemicals at U.S. sources), http://www.scorecard.org.

HAZARDOUS WASTE TREATMENT

Federal Remedial Technologies Roundtable, Hazardous Waste Clean-Up Information ("CLU-IN"), http://www.clu-in.org.

HEAVY METALS

U.S. Environmental Protection Agency, Office of Pollution Prevention and Toxics, http://www.epa.gov/opptintr.

HIGH-LEVEL RADIOACTIVE WASTES

U.S. Nuclear Regulatory Commission, Radioactive Waste Page, http://www.nrc.gov/NRC/radwaste.

U.S. Environmental Protection Agency, Mixed-Waste Homepage, http://www.epa.gov/radiation/mixed-waste.

HURRICANES

National Hurricane Center, http://www.nhc.noaa.gov.

HYDROELECTRIC POWER

U.S. Bureau of Reclamation Hydropower Information, http://www.usbr.gov/power/edu/edu.htm.

U.S. Geological Survey, http://wwwga.usgs.gov/edu/hybiggest.html.

HYDROGEN

National Renewable Energy Laboratory, http://www.nrel.gov/lab/pao/hydrogen.html.

EnviroSource, Hydrogen InfoNet, http:///www.eren.doe.gov/hydrogen/infonet.html.

INTERNATIONAL ATOMIC ENERGY AGENCY

Agency, http://www.iaea.org.

Managing Radioactive Waste Fact Sheet, http://www.iaea.org/worldatom/inforesource/factsheets/manradwa.html.

INTERNATIONAL COUNCIL FOR LOCAL ENVIRONMENTAL INITIATIVES

Homepage, http://www.iclei.org.

INTERNATIONAL REGISTER OF POTENTIALLY TOXIC CHEMICALS

Homepage, http://www.unep.org/unep/program/hhwb/chemical/irptc/home.htm.

INTERNATIONAL WHALING COMMISSION

Homepage, http://www.ourworld.compuserve.com/homepages/iwcoffice.

INVERTEBRATES: THREATENED AND ENDANGERED

U.S. Fish and Wildlife Service, Species List of Endangered and Threatened Wildlife, http://endangered.fws.gov/

LANDSAT AND SATELLITE IMAGES

Earthshots, Satellite Images of Environmental Change, http://www.usgs.gov/Earthshots/.

Landsat Gateway, http://landsat.gsfc.nasa.gov/main.htm.

LEAD

National Lead Information Center's Clearinghouse, 1-800-424-LEAD, http://www.epa.gov/lead/.

LEOPARDS

U.S. Fish and Wildlife Service, Species List of Endangered and Threatened Wildlife, http://www.fws.gov/r9endspp/lsppinfo.html.

LITTER

Keep America Beautiful, http://www.kab.org.

MAMMALS

U.S. Fish and Wildlife Service, Vertebrate Animals, http://www.fws.gov/r9endspp/lsppinfo.html.

MANATEES

Save the Manatees, http://www.savethemanatee. org.

Sea World, Manatees, http://www.seaworld.org/manatee/sciclassman.html.

MARSHES

Environmental Protection Agency, Office of Wetlands, Oceans, Watersheds, http://www.epa.gov/owow/wetlands/wetland2.html.

North American Waterfowl and Wetlands Office, http://www.fws.gov/r9nawwo.

North American Wetlands Conservation Act, http://www.fws.gov/r9nawwo/nawcahp.html.

North American Wetlands Conservation Council, http://www.fws.gov/r9nawwo/nawcc.html.

Wetlands, wetlands-hotline@epamail.epa.gov.

MATERIAL SAFETY DATA SHEET

Toxic chemicals, http://www.siri.org/msds; http://www.ilpi.com/mads/index.html.

MENDES, CHICO

Chico Mendes, http://www.edf.org/chico.

NATURAL DISASTERS

Building Safer Structures, http://quake.wr.usgs. gov/QUAKES/FactSheets/SaferStructures/.

Center for Integration of Natural Disaster Information, http://cindi.usgs.gov/events/.

Earthquakes, http://quake.wr.usgs.gov/; http://geology.usgs.gov/quake.html. For the latest earthquake information http://quake.wr.usgs.gov/QUAKES/CURRENT/current.html

National Hurricane Center, http://www.nhc.noaa.gov.

U.S. Geological Survey, http://geology.usgs.gov/whatsnew.html.

NATIONAL MARINE FISHERIES

History of National Marine Fisheries Service, http://www.wh.whoi.edu/125th/history/century.html.

National Marine Fisheries, http://kingfish.ssp.nmfs.gov.

NOAA Fisheries, http://www.nmfs.gov/.

NATIONAL OCEAN AND ATMOSPHERIC ADMINISTRATION (NOAA)

Climate forecasting, http://www.cdc.noaa.gov/ Seasonal/.

El Niño Theme Page, http://www.pmel.noaa.gov/toga-tao/el-nino/nino-home-low.html.

Homepage, http://www.noaa.gov/.

Recover Protected Species, http://www.noaa.gov/nmfs/recover.html.

Safe Navigation Page, http://anchor.ncd.noaa.gov/psn/psn.htm.

NATIONAL WEATHER SERVICE

Homepage, http://www.nws.noaa.gov.

NATIONAL WILDLIFE REFUGE SYSTEM

Homepage, http://refuges.fws.gov/NWRSHomePage.html.

NATURAL GAS

American Gas Association, http://www.aga.org.

Oil and Gas Journal Online, http://www.ogjonline.com.

U.S. Department of Energy, Energy Information Administration, http://www.eia.doe.gov.

U.S. Department of Energy, Office of Fossil Energy, http://www.fe.doe.gov.

U.S. Geological Survey Energy, Resources Program, http://energy.usgs.gov/index.html.

NOISE POLLUTION

Noise Pollution Clearinghouse, http://www. nonoise.org.

NONPOINT SOURCES

Nonpoint Source Pollution Control Program, http://www.epa.gov/OWOW/NPS/whatudo.html; http://www.epa.gov/OWOW/ NPS/.

NUCLEAR ENERGY AND NUCLEAR REACTORS

American Nuclear Society, http://www.ans.org.

Nuclear Energy Institute, http://www.nei.org.

Nuclear Information and Resource Service, http://www.nirs.org.

U.S. Department of Energy, Office of Nuclear Energy, Science and Technology, http://www.ne.doe.gov.

U.S. Nuclear Regulatory Commission, http://www.nrc.gov.

NUCLEAR WASTE POLICY ACT

American Nuclear Society, http://www.ans.org.

Nuclear Energy Institute, http://www.nei.org.

NUCLEAR WASTE SITES

Hazard Ranking System, http://www.epa.gov/ superfund/programs/npl_hrs/hrsint.htm.

National Research Council, Board on Radioactive Waste Management, http://www4.nas.edu/brwm/brwm-res.nsf.

Superfund, http://www.pin.org/superguide.htm; http://www.epa.gov/superfund.

U.S. Department of Energy, Office of Civilian Radioactive Waste Management, http://www.rw.doe.gov.

U.S. Environmental Protection Agency, Mixed-Waste Homepage, http://www.epa.gov/radiation/mixed-waste.

U.S. Nuclear Regulatory Commission, Radioactive Waste Page, http://www.nrc.gov/ NRC/radwaste.

OCCUPATIONAL SAFETY AND HEALTH ACT (OSHA)

OSHA Homepage, http://www.osha.gov.

OCEAN THERMAL ENERGY CONVERSION (OTEC)

National Renewable Energy Laboratory

1617 Cole Boulevard
Golden, CO 80401
Website: http:llnrelinfo.nrel.gov.

Natural Energy Laboratory of Hawaii, http://bigisland.com/nelha/index.html.

OCEANS

National Oceanographic and Atmospheric Administration, http://www.noaa.gov/.

Safe Ocean Navigation Page, http://anchor.ncd.noaa.gov/psn/psn.htm.

OFFICE OF SURFACE MINING

Office of Surface Mining, http://www.osmre.gov.

Appalachian Clean Streams Initiative, majordomo@osmre.gov.

OLD-GROWTH FORESTS

Greenpeace International, Forests, http://www.greenpeace.org/~forests.

World Resources Institute, Forest Frontiers Initiative, http://www.wri.org/ffi.

OLMSTEAD, FREDERICK LAW

Homepage, http://fredericklawolmsted.com.

ORGANIZATION OF PETROLEUM EXPORTING COUNTRIES (OPRC)

Homepage, http://www.opec.org.

OVERFISHING

Information and data statistics, http://www.nmfs. gov.

National Aeronautics and Space Administration, Ocean Planet, http://seawifs.gsfc.nasa.gov/OCEAN_PLANET/HTML/peril_overfishing.html.

National Marine Fisheries Service, http://www. nmfs.gov.

NOAA, http://www.noaa.gov.

United Nations Food and Agricultural Organization, http://www.fao.org.

United Nations Food and Agriculture Organization Fisheries, http://www.fao.org/.

United Nations System, http://www.unsystem.org.

OZONE-RELATED ISSUES

Environmental Protection Agency, science of ozone depletion, http://www.epa.gov/ozone/science/.

NOAA, Commonly Asked Questions about Ozone, www.publicaffairs.noaa.gov/grounders/ozo1.html.

NOAA, Network for the Detection of Stratospheric Change, www.noaa.gov.

PARROTS

Online Book of Parrots, http://www.ub.tu-clausthal.dep/p_welcome.html.

World Parrot Trust, http://www.worldparrottrust.org.

World Wildlife Fund, http:www.panda.org.

PESTICIDES

Toxics and Pesticides, http://www.epa.gov/oppfead1/work_saf/.

Pesticides in the Atmosphere, http://ca.water.usgs.gov/pnsp/atmos.

PETERSON, ROGER TORY

Roger Tory Peterson Institute of Natural History,

311 Curtis Street
Jamestown, NY 14701
Website: http://www.rtpi.org/info/rtp.htm.

PETROLEUM

American Petroleum Institute, http://www.api.org.

Petroleum Information, http://www.petroleuminformation.com.

Oil and Gas Journal Online, http://www. ogjonline.com.

U.S. Department of Energy, Energy Information Administration, http://www.eia.doe.gov.

U.S. Department of Energy, Office of Fossil Energy, http://www.fe.doe.gov.

U.S. Geological Survey Energy Resources Program, http://energy.usgs.gov/index.html.

U.S. Geological Survey Fact Sheet FS-145-97, "Changing Perceptions of World Oil and Gas Resources as Shown by Recent USGS Petroleum Assessments," http://greenwood.cr.usgs.gov/pub/fact-sheets/fs-0145-97/fs-0145-97.html.

PLUTONIUM

U.S. Nuclear Regulatory Commission, Radioactive Waste Page, http://www.nrc.gov/NRC/radwaste.

RADIATION AND RADIOACTIVE WASTES

International Atomic Energy Agency, "Managing Radioactive Waste" Fact Sheet, http://www.iaea.org/worldatom/inforesource/factsheets/manradwa.html.

National Research Council, Board on Radioactive Waste Management, http://www4.nas.edu/brwm/brwm-res.nsf.

U.S. Department of Energy, Office of Civilian Radioactive Waste Management, http://www. rw.doe.gov.

U.S. Environmental Protection Agency, Mixed-Waste Homepage, http://www.epa.gov/radiation/mixed-waste.

U.S. Nuclear Regulatory Commission, Radioactive Waste Page, http://www.nrc.gov/NRC/radwaste.

RADON

Radon in Earth, Air, and Water, http://sedwww.cr.usgs.gov:8080/radon/radonhome.html.

RAIN FORESTS

Greenpeace International, forests, http://www.greenpeace.org/~forests.

Rainforest Action Network (RAN)

President Randy Hayes
221 Pine Street Suite 500
San Francisco, CA 94104
Telephone: (415) 398-4404
Website: http://www.ran.org

Rainforest Alliance (RA)

65 Bleeker Street
New York, NY 10012
Website: http://www.rainforest-alliance.org

U.S. Forest Service, http://www.fs.fed.us.

World Wildlife Fund (Worldwide Fund for Nature), Forests for Life Campaign, http://www.panda.org/forests4life.

RESOURCE CONSERVATION AND RECOVERY ACT

Homepage, http://www.epa.gov/epaoswer/hotline.

SALMON

National Marine Fisheries Service, http://www.nwr.noaa.gov/1salmon/salmesa/index.htm.
NOAA Fisheries, http://www.nmfs.gov/.

SALT MARSHES

National Wetlands Research Center, http://www.nwrc.usgs.gov/educ_out.html.

USGS Coastal and Marine Geology, http://marine.usgs.gov/.

SANITARY LANDFILLS

Solid waste management, http://web.mit.edu/urbanupgrading/urban environment/

Landfills - Solid and Hazardous Waste and Groundwater Quality Protection, http://www.gfredlee.com/plandfil2.htm

SIBERIA

Siberia, http://www.cnit.nsk.su/univer/english/siberia.htm.

SOLAR ENERGY

American Solar Energy Society

2400 Central Avenue, Suite G-1
Boulder, CO 80301.
Website: http://www.soton.ac.uk/~solar/.

Solar Energy Industries Association

122 C Street, NW, 4th Floor
Washington, D.C. 20001.
Website: http://www.seia.org/main.htm.

U.S. Department of Energy, Photovoltaic Program, http://www.eren.doe.gov/pv/text_frameset.html.

SOLAR POND

Department of Mechanical and Industrial Engineering

University of Texas at El Paso
El Paso, TX 79968.
E-mail: aswift@cs.utep.edu.

SPENT FUEL

Environmental Protection Agency, www.ntp.doe.gov, www.rw.doe.gov/pages/resource/facts/transfct.htm.

SUPERFUND

Environmental Protection Agency, http://www.epa.gov/epaoswer/hotline.

Recycled Superfund sites, http://www.epa.gov/superfund/programs/recycle/index.htm.

Superfund Information, http://www.epa.gov/superfund.

U.S. EPA Superfund Program Homepage, Website: http://www.epa.gov/superfund/index.htm.

TENNESSEE VALLEY AUTHORITY

Homepage, http://www.tva.gov.

THOREAU, HENRY

Website: http://www.walden.org.

TOXIC CHEMICALS

Environmental Defense Fund, http://www.scorecard.org.

U.S. Department of Health and Human Services, Agency for Toxic Substances and Disease Registry (ASTDR), http://www.atsdr.cdc.gov/

U.S. Environmental Protection Agency, Integrated Risk Information System (IRIS), http://www. siri.org/msds; http://www.ilpi.com/mads/index.html.

U.S. Occupation Health and Safety Administration, http://www.toxicsubstances/index.html.

TOXIC RELEASE INVENTORY

Environmental Defense Fund, http://www.scorecard.org.
Environmental Protection Agency, http://www.epa.gov.

Teach with Databases, Toxic Release Inventory, http://www.nsta.org/pubs/special/pb143x01.htm.

TOXIC WASTE

Environmental Defense Fund, http://www.scorecard.org.

Institute for Global Communications, http://www.igc.org/igc/issues/tw/.

TRADE RECORDS ANALYSIS OF FLORA AND FAUNA IN COMMERCE (TRAFFIC)

Homepage, http://www.traffic.org/about/.

URBAN FORESTS

American Forests, http://www.amfor.org.

TreeLink, http://www.treelink.org.

VERTEBRATES

U.S. Fish and Wildlife Service, Species List of Endangered and Threatened Wildlife, http://www.fws.gov/r9endspp/lsppinfo.html.

VICUNA

U.S. Fish and Wildlife Service, Species List of Endangered and Threatened Wildlife, http://endangered.fws.gov

VITRIFICATION

U.S. Department of Energy, http://www.em.doe.gov/fs/fs3m.html.

VOLCANOES

USGS, Volcanoes in the Learning Web, http://www.usgs.gov/education/learnweb/volcano/index.html.

Volcano Hazards, http://volcanoes.usgs.gov/.

WATER CONSERVATION AND POLLUTION

Early History of the Clean Water Act, http://epa.gov/history/topics.

Environmental Protection Agency, Office of Wetlands, Oceans, Watersheds for Nonpoint Source information, http://www.epa.gov/owow/wetlands/wetland2.html; http://www.epa.gov/swerosps/ej/.

U.S. Geological Survey, Water Resources of the United States, National Groundwater Association Homepage, http://www.h2o-ngwa.org.

Water Resources Information, http://water.usgs.gov/.

Water Use Data, http://water.usgs.gov/public/watuse/.

WETLANDS

National Wetlands Research Center, http://www.nwrc.usgs.gov/educ_out.html.

Ramsar Convention on Wetlands (International), http://www2.iucn.org/themes/ramsar/.

Ramsar List of Wetlands of International Importance, http://ramsar.org/key_sitelist.htm.

WHALES

Institute of Cetacean Research (ICR), http://www.whalesci.org.

U.S. Fish and Wildlife Service, Species List of Endangered and Threatened Wildlife, http://www.fws.gov/r9endspp/lsppinfo.html; http://www.highnorth.no/iceland/th-in-to.htm; http://greenpeace.org/.

WILDERNESS

U.S. Forest Service, *Roadless Area Review and Evaluation*, http://www.fs.fed.us.

Wilderness Society, http://www.wilderness.org/newsroom/factsheets.htm.

WILDLIFE REFUGES

Conservation International, http://www.conservation.org.

Nature Conservancy, http://www.tnc.org.

U.S. Fish and Wildlife Service, National Wildlife Refuge System, http://refuges.fws.gov.

World Conservation Union/International Union for the Conservation of Nature, http://www.iucn.org.

WIND ENERGY

American Wind Energy Association

122 C Street NW, 4th Floor
Washington, D.C. 20001
Telephone: (202) 383-2500.
E-mail: awea@mcimail.com.
Website: http://www.awea.org.

Center for Renewable Energy and Sustainable Technology (CREST)

Solar Energy Research and Education Foundation
777 North Capitol Street NE, Suite 805
Washington, D.C. 20002
Website: http://solstice.crest.org/.

WOLVES

U.S. Fish and Wildlife Service, http://www.fws.gov/.
U.S. Fish and Wildlife Service, Species List of Endangered and Threatened Wildlife, http://endangered.fws.gov/.

World Wildlife Fund

1250 24th Street, NW
Washington, D.C. 20037
Telephone: 1-800-225-5993
Website: http://www.worldwildlife.org/.

WORLD HEALTH ORGANIZATION

Homepage, http://www.who.int.

WORLD WILDLIFE FUND

1250 24th Street, NW
Washington, D.C. 20037
Telephone: 1-800-225-5993
Website: http://www.wwf.org/.

YUCCA MOUNTAIN PROJECT

Homepage, http://www.ymp.gov/.

ZEBRAS

U.S. Fish and Wildlife Service, Species List of Endangered and Threatened Wildlife, http://endangered.fws.gov/.

ZOOS

Bronx Zoo, http://www.bronxzoo.com/.
San Diego Zoo, http://www.sandiegozoo.org/.

Appendix D: Environmental Organizations

Action for Animals

P.O. Box 17702
Austin, TX 78760
Telephone: (512) 416-1617
Fax: (512) 445-3454
Website: http://www.envirolink.org/

African Wildlife Foundation (AWF)

1400 Sixteenth Street, NW, Suite 120
Washington, D.C. 20036
Telephone: (202) 939-3333
Fax: (202) 939-3332
Website: http://www.awf.org/home.html

Agency for Toxic Substances and Diseases, Registry Division of Toxicology (ATSDR)

1600 Clifton Road
NE Mailstop E-29
Atlanta, GA 30333
Telephone: (888) 42-ATSDR or (888) 422-8737
E-mail: ATSDRIC@cdc.gov
Website: http://www.atsdr.cdc.gov/contacts.html

Alaska Forum for Environmental Responsibility

P.O. Box 188
Valdez, AK 99686
Telephone: (907) 835-5460
Fax: (907) 835-5410
Website: http://www.accessone.com/~afersea

American Conifer Society (ACS)

P.O. Box 360
Keswick, VA 22947-0360
Telephone: (804) 984-3660
Fax: (804) 984-3660
E-mail: ACSconifer@aol.com
Website: http://www.pacificrim.net/~bydesign/acs.html

American Forests

P.O. Box 2000
Washington, D.C. 20013
Telephone: (202) 955-4500
Website: http://www.americanforests.org

American Nuclear Society

555 North Kensington Avenue
La Grange Park, IL 60525
Telephone: (708) 352-6611
Fax: (708) 352-0499
E-mail: NUCLEUS@ans.org
Website: http://www.ans.org

American Oceans Campaign

201 Massachusetts Avenue NE, Suite C-3
Washington, D.C. 20002
Telephone: (202) 544-3526
Fax: (202) 544-5625
E-mail: aocdc@wizard.net
Website: http://www.americanoceans.org

American Rivers

1025 Vermont Avenue NW, Suite 720
Washington, D.C. 20005
Telephone: (202) 347-7500
Fax: (202) 347-9240
E-mail: amrivers@amrivers.org
Website: http://www.amrivers.org

American Society for Horticultural Science (ASHS)

600 Cameron Street
Alexandria, VA 22314-2562

Telephone: (703) 836-4606
Fax: (703) 836-2024
E-mail: webmaster@ashs.org
Website: http://www.ashs.org

American Society for the Prevention of Cruelty to Animals (ASPCA)

424 East Ninety-second Street
New York, NY 10128
Telephone: (212) 876-7700
Website: http://www.aspca.org

American Solar Energy Society

2400 Central Avenue, Suite G-1
Boulder, CO 80301
Telephone: (303) 443-3130
Fax: (303) 443-3212
E-mail: ases@ases.org
Website: http://www.ases.org
Publication: *Solar Today*

American Wind Energy Association

122 C Street NW, Fourth Floor
Washington, D.C. 20001
Telephone: (202) 383-2500
E-mail: awea@mcimail.com
Website: http://www.awea.org

Animal Legal Defense Fund (ALDF)

127 Fourth Street
Petaluma, CA 94952
Telephone: (707) 769-7771
Fax: (707) 769-0785
E-mail: info@aldf.org
Website: http://www.aldf.org

Animal Rights Network

P.O. Box 25881
Baltimore, MD 21224
Telephone: (410) 675-4566
Fax: (410) 675-0066
Website: http://www.envirolink.org/arrs/aa/index.html
Publication: *Animals' AGENDA*, a bimonthly magazine

Baron's Haven Freehold

104 South Main Street
Cadiz, OH 43907
Telephone: (740) 942-8405
Website: http://bhfi.1st.net

Biodiversity Support Program (BSP)

1250 North Twenty-fourth Street NW, Suite 600
Washington, D.C. 20037
Telephone: (202) 778-9681
Fax: (202) 861-8324
Website: http://www.BSPonline.org

Biosfera

Pres. Vargas 435, Suites 1104 and 1105
Rio de Janeiro, RJ 20077-900
Brazil

Birds of Prey Foundation

2290 South 104th Street
Broomfield, CO 80020
Telephone: (303) 460-0674
Fax: (303) 666-1050
E-mail: raptor@birds-of-prey.org
Website: http://www.birds-of-prey.org

Build the Earth

3818 Surfwood Road
Malibu, CA 90265
Telephone: (310) 454-0963

Center for Conversion and Research of Endangered Wildlife (CREW)

Cincinnati Zoo and Botanical Garden
3400 Vine Street
Cincinnati, OH 45220
E-mail: terri.roth@cincyzoo.org

Center for Marine Conservation

1725 DeSales Street SW, Suite 600
Washington, D.C. 20036
Telephone: (202) 429-5609
Fax: (202) 872-0619
E-mail: cmc@dccmc.org
Website: http://www.cmc-ocean.org

Centers for Disease Control (CDC)

1600 Clifton Rd.
Atlanta, GA 30333

Telephone: (800) 311-3435
Website: http://www.cdc.gov

Cheetah Conservation Fund (CCF)

P.O. Box 1380
Ojai, CA 93024
Telephone: (805) 640-0390
Fax: (815) 640-0230
E-mail: info@cheetah.org
Website: http://www.cheetah.org

Clean Air Council (CAC)

135 South Nineteenth Street, Suite 300
Philadelphia, PA 19103
Telephone: (888) 567-7796
Website: http://www.libertynet.org/~cleanair/

Coalition for Economically Responsible Economies (CERES)

11 Arlington Street, Sixth Floor
Boston, MA 02116-3411
Telephone: (617) 247-0700
Fax: (617) 267-5400
Website: http://www.ceres.org

Conservation International

1015 Eighteenth Street NW Suite 1000
Washington, D.C. 20036
Telephone: (202) 429-5660
Website: http://www.conservation.org/
Publication: *Orion Nature Quarterly*

Convention on International Trade in Endangered Species of Wild Fauna and Flora (CITES)

CITES Secretariat
International Environment House, 15, chemin des Anémones, CH-1219
Châtelaine-Geneva, Switzerland
E-mail: cites@unep.ch
Website: http://www.cites.org/index.shtml

Council for Responsible Genetics

5 Upland Road, Suite 3
Cambridge, MA 02140
Website: http://www.gene-watch.org

Cousteau Society

870 Greenbriar Circle, Suite 402
Chesapeake, VA 23320
Telephone: (804) 523-9335
E-mail: cousteau@infi.net
Website: http://www.cousteausociety.org/
Publication: *Calypso Log*

Defenders of Wildlife

1101 Fourteenth Street NW, Room 1400
Washington, D.C. 20005
Telephone: (800) 441-4395
Website: http://www.Defenders.org
Publication: *Defenders*, a quarterly magazine

Dian Fossey Gorilla Fund International

800 Cherokee Avenue SE
Atlanta, GA 30315-1440
Telephone: (800) 851-0203
Fax: (404) 624-5999
E-mail: 2help@gorillafund.org
Website: http://www.gorillafund.org/000_core_frmset.html

Earth Day Network

1616 P Street NW
Suite 200
Washington, D.C. 20036
E-mail: earthday@earthday.net
Website: http://www.earthday.net

Earth Island Institute (EII)

300 Broadway, Suite 28
San Francisco, CA 94133
Telephone: (415) 788-3666
Fax: (415) 788-7324
Website: http://www.earthisland.org/abouteii/abouteii.html
Publication: *Earth Island Journal*, a quarterly magazine

Earth, Pulp, and Paper

P.O. Box 64
Leggett, CA 95585
Telephone: (707) 925-6494
E-mail: tree@tree.org
Website: http://www.tree.org/epp.htm

EarthFirst! (EF!)

P.O. Box 5176
Missoula, MT 59806
Website: http://www.webdirectory.com/General_Environmental_Interest/Earth_First_/

Earthwatch Institute

In United States and Canada
3 Clocktower Place, Suite 100
Box 75
Maynard, MA 01754
Telephone: (800) 776-0188 or (617) 926-8200
Fax: (617) 926-8532
In Europe
57 Woodstock Road
Oxford OX2 6HJ, United Kingdom
E-mail: info@uk.earthwatch.org
Website: http://www.earthwatch.org

EcoCorps

1585 A Folsom Avenue
San Francisco, CA 94103
Telephone: (415) 522-1680
Fax: (415) 626-1510
E-mail: eathvoice@ecocorps.org
Website: http://www.owplaza.com/eco

Ecotourism Society

P.O. Box 755
North Bennington, VT 05257
Telephone: (802) 447-2121
Fax: (802) 447-2122
E-mail: ecomail@ecotourism.org
Website: http://www.ecotoursim.org

E. F. Schumacher Society

140 Jug End Road
Great Barrington, MA 01230
Telephone: (413) 528-1737
E-mail: efssociety@aol.com
Website: http://members.aol.com/efssociety/index.html

Electric Vehicle Association of the Americas

701 Pennsylvania Avenue NW, Fourth Floor
Washington, D.C. 20004
Telephone: (202) 508-5995
Fax: (202) 508-5924
Website: http://www.evaa.org

Environmental Defense Fund (EDF)

257 Park Avenue South
New York, NY 10010
Telephone: (800) 684-3322
Fax: (212) 505-2375
E-mail (for general questions and information): Contact@environmentaldefense.org
Website: http://www.edf.org
Publication: *Nature Journal*, a monthly magazine

Exotic Cat Refuge and Wildlife Orphanage

Route 3, Box 96A
Kirbyville, TX 75956
Telephone: (409) 423-4847

Federal Emergency and Management Agency (FEMA)

500 C Street SW
Washington, D.C. 20472
Website: http://www.fema.gov

Friends of the Earth (FOE)

1025 Vermont Avenue NW, Suite 300
Washington, D.C. 20005-6303
Telephone: (202) 783-7400
Fax: (202) 783-0444
E-mail: foe@foe.org
Website: http://www.foe.org

Green Seal

1001 Connecticut Avenue NW, Suite 827
Washington, D.C. 20036-5525
Telephone: (202) 872-6400
Fax: (202) 872-4324
Website: http://www.greenseal.org

Greenpeace USA

1436 U Street NW
Washington, D.C. 20009
Telephone: (202) 462-1177
Website: http://www.greenpeaceusa.org/
Publication: *Greenpeace Magazine*

Hawkwatch International

P.O. Box 660
Salt Lake City, UT 84110
Telephone: (801) 524-8511
E-mail: hawkwatch@charitiesusa.com
Website: http://www.vpp.com/HawkWatch

Humane Society of the United States (HSUS)

2100 L Street NW
Washington, D.C. 20037
Website: http://www.hsus.org
Publications: *All Animals*, a quarterly magazine

International Atomic Energy Commission

P.O. Box 100
Wagramer Strasse 5
A-1400, Vienna, Austria
E-mail: Official.Mail@iaea.org
Website: http://www.iaea.org

International Council for Local Environmental Initiatives (ICLEI)

World Secretariat
16th Floor, West Tower, City Hall
Toronto, M5H 2N2, Canada
Fax: (416) 392-1478
Email: iclei@iclei.org
Website: http://www.iclei.org

International Rhino Foundation (IRF)

14000 International Road
Cumberland Ohio 43732
E-mail: IrhinoF@aol.com
Website: http://www.rhinos-irf.org

International Whaling Commission (IWC)

The Red House
135 Station Road
Impington, Cambridge CB4 9NP, United Kingdom
E-mail: iwc@iwcoffice.org
Website: http://ourworld.compuserve.com/homepages/iwcoffice

International Wolf Center

1396 Highway 169
Ely, MN 55731-8129
Telephone: (218) 365-4695
Fax: (218) 365-3318
Website: http://www.wolf.org

Jane Goodall Institute (JGI)

P.O. Box 14890
Silver Spring, MD 20911-4890
Telephone: (301) 565-0086
Fax: (301) 565-3188
E-mail: JGIinformation@janegoodall.org

Keep America Beautiful

1010 Washington Boulevard
Stamford, CT 06901
Telephone: (203) 323-8987
Fax: (203) 325-9199
E-mail: info@kab.org

League of Conservation Voters

1707 L Street, NW, Suite 750
Washington, D.C. 20036
Telephone: (202) 785-8683
Fax: (202) 835-0491
E-mail: lcv@lcv.org
Website: http://www.lcv.org

Mountain Lion Foundation (MLF)

P.O. Box 1896
Sacramento, CA 95812
Telephone: (916) 442-2666
E-mail: MLF@moutainlion.org
Website: http://www.mountainlion.org

National Alliance of River, Sound, and Bay Keepers

P.O. Box 130
Garrison, NY 10524
Telephone: (800) 217-4837
E-mail: keepers@keeper.org
Website: http://www.keeper.org

National Anti-Vivisection Society (NAVS)

53 West Jackson Street, Suite 1552
Chicago, IL 60604
Telephone: (800) 888-NAVS
E-mail: navs@navs.org
Website: http://www.navs.org

National Arbor Day Foundation

100 Arbor Avenue
Nebraska City, NE 68410
Telephone: (402) 474-5655
Website: http://www.arborday.org
Publication: *Arbor Day*, a bimonthly magazine

National Audubon Society (NAS)

700 Broadway
New York, NY 10003
Telephone: (212) 979-3000
Website: http://www.audubon.org
Publication: *Audubon*, a bimonthly magazine

National Center for Environmental Health

Mail Stop F-29
4770 Buford Highway NE
Atlanta, GA 30341-3724
Telephone NCEH Health Line: (888) 232-6789
Website: http://www.cdc.gov/nceh/ncehhome.htm

National Parks and Conservation Association (NPCA)

1015 Thirty-first Street NW
Washington, D.C. 20007
Telephone: (202) 944-8530; (800) NAT-PARK
E-mail: npca@npca.org
Website: http://www.npca.org
Publication: *National Parks*, a bimonthly magazine

National Wildlife Federation (NWF)

8925 Leesburg Pike
Vienna, VA 22184-0001
Telephone: (800) 822-9919
Website: http://www.nwf.org
Publication: *National Wildlife*, a bimonthly magazine

Natural Resources Defense Council (NRDC)

40 West Twentieth Street
New York, NY 10011
Website: http://www.nrdc.org
Publications: *Amiscus Journal*, a quarterly magazine

Nature Conservancy (TNC)

1815 North Lynn Street
Arlington, VA 22209
Telephone: (703) 841-5300
Fax: (703) 841-1283
Website: http://www.tnc.org
Publication: *Nature Conservancy*, a magazine

Noise Pollution Clearinghouse

P.O. Box 1137
Montpelier, VT 05601-1137
Telephone: (888) 200-8332
Website: http://www.nonoise.org

North Sea Commission

Business and Development Office
Skottenborg 26, DK-8800 Viborg, Denmark
Website: http:\\www.northsea.org

People for Animal Rights

P.O. Box 8707
Kansas City, MO 64114
Telephone: (816) 767-1199
E-mail: parinfo@envirolink.org
Website: http://www.parkc.org

People for the Ethical Treatment of Animals (PETA)

501 Front Street
Norfolk, VA 23510
Telephone: (757) 622-PETA
Fax: (757) 622-0457
Website: http://www.peta-online.org/

Orangutan Foundation International

822 South Wellesley Avenue
Los Angeles, CA 90049
Telephone: (800) ORANGUTAN
Fax: (310) 207-1556
E-mail: ofi@orangutan.org
Website: http://www.ns.net/orangutan

Ozone Action

1700 Connecticut Avenue NW, Third Floor
Washington, D.C. 20009
Telephone: (202) 265-6738

E-mail: cantando@essential.org
Website: www.ozone.org

Peregrine Fund

566 West Flying Hawk Lane
Boise, ID 83709
Telephone: (208) 362-3716
Fax: (208) 362-2376
E-mail: tpf@peregrinefund.org
Website: http://www.peregrinefund.org

Rachel Carson Council

8940 Jones Mill Road
Chevy Chase, MD 20815
Telephone: (301) 652-1877
E-mail: rccouncil@aol.com
Website: http://members.aol.com/rccouncil/ourpage

Rainforest Action Network

221 Pine Street, Suite 500
San Francisco, CA 94104-2740
Telephone: (415) 398-4404
Fax: (415) 398-2732
E-mail: rainforest@ran.org
Website: http://www.ran.org

Range Watch

45661 Poso Park Drive
Posey, CA 93260
Telephone: (805) 536-8668
E-mail: rangewatch@aol.com
Website: http://www.rangewatch.org

Raptor Resource Project

2580 310th Street
Ridgeway, IA 52165
E-mail: rrp@salamander.com
Website: http://www.salamander.com~rpp

Reef Relief

201 William Street
Key West, FL 33041
Telephone: (305) 294-3100
Fax: (305) 923-9515
E-mail: reef@bellsouth.net
Website: http://www.reefrelief.org

ReefKeeper International

2809 Bird Avenue, Suite 162
Miami, FL 33133
Telephone: (305) 358-4600
Fax: (305) 358-3030
E-mail: reefkeeper@reefkeeper.org
Website: http://www.reefkeeper.org

Renewable Energy Policy Project-Center for Renewable Energy and Sustainable Technology (REPP-CREST)

National Headquarters
1612 K Street, NW, Suite 202
Washington, D.C. 20006
Website: http://www.solstice.crest.org

Resources for the Future (RFF)

1616 P Street NW
Washington, D.C. 20036
Telephone: (202) 328-5000
Fax: (202) 939-3460
E-mail: info@rff.org
Website: http://www.rff.org

Roger Tory Peterson Institute

311 Curtis Street
Jamestown, NY 14701
Telephone: (716) 665-2473
E-mail: webmaster@rtpi.org

Sierra Club

85 Second Street, Second Floor
San Francisco, CA 94105
Telephone: (415) 977-5630
Fax: (415) 977-5799
E-mail (general information): information@sierraclub.org
Website: http://www.Sierraclub.org
Publication: *Sierra*, a bimonthly magazine

Smithsonian Institution Conservation & Research Center (CRC)

Website: http://www.si.edu/crc/brochure/index.htm

Society of American Foresters

5400 Grosvenor Lane
Bethesda, MD 20814

Telephone: (301) 897-8720
Fax: (301) 897-3690
E-mail: safweb@safnet.org
Website: http://www.safnet.org

Surfrider Foundation USA

122 South El Camino Real, Suite 67
San Clemente, CA 92672
Telephone: (949) 492-8170
Fax: (949) 492-8142
Website: http://www.surfrider.org

Union of Concerned Scientists

National Headquarters
2 Brattle Square
Cambridge, MA 02238
Telephone: (617) 547-5552
E-mail: ucs@ucsusa.org
Website: http://www.ucsusa.org
Publications: *Nucleus*, a quarterly magazine; *Earthwise*, a quarterly newsletter

United Nations Environment Programme (Regional)

2 United Nations Plaza
NY, NY 10017
Telephone: (212) 963-8138
Website: http://www.unep.org

United Nations Food and Agriculture Organization (FAO)

Website: http://www.fao.org
Liaison office with North America
Suite 300, 2175 K Street NW, Washington D.C. 20437-0001

United Nations Man and the Biosphere Programme (UNMAB)

U.S. MAB Secretariat, OES/ETC/MAB
Department of State
Washington, D.C. 20522-4401
Website: http://www.mabnet.org

U.S. Department of Agriculture (USDA)

14th Street and Independence Avenue., SW,
Washington, D.C. 20250
Website: http://www.usda.gov

U.S. Department of Energy (DOE)

Forrestal Building
1000 Independence Avenue, SW,
Washington, D.C. 20585
Website: http://www.doe.gov

U.S. Environmental Protection Agency (EPA)

401 M Street SW
Washington, D.C. 20460
Website: http://www.epa.gov

U.S. Fish and Wildlife Service (FWS)

1849 C Street NW
Washington, D.C. 20240
Telephone: (202) 208-5634
Website: http://www.fws.org

U.S. Geological Survey (USGS)

U.S. Dept. of Interior
1849 C Street, NW
Washington, D.C. 20240
Website: http://www.usgs.gov

U.S. National Park Service (NPS)

U.S. Dept. of Interior
1849 C Street, NW
Washington, D.C. 20240
Website: http://www.nps.gov

U.S. Nuclear Regulatory Commission (NRC)

One White Flint North
11555 Rockville Pike
Rockville, Maryland 20852
Website: http://www.nrc.gov

Wilderness Society

900 Seventeenth Street NW
Washington, D.C. 20006-2506
Telephone: (800) THE-WILD
Website: www.wilderness.org

Wildlands Project (TWP)

1955 West Grant Road, Suite 145
Tucson, AZ 85745
Telephone: (520) 884-0875
Fax: (520) 884-0962

E-mail: information@twp.org
Website: http://www.twp.org

World Conservation Monitoring Centre (WCMC)

219 Huntington Road
Cambridge CB3 ODL, United Kingdom
E-mail: info@wcmc.org.uk
Website: http://www.wcmc.org.uk

World Conservation Union (IUCN)

1630 Connecticut Avenue NW, Third Floor
Washington, D.C. 20009-1053
Telephone: (202) 387-4826
Fax: (202) 387-4823
E-mail: postmaster@iucnus.org
Website: http://www.iucn.org

World Health Organization (WHO)

Avenue Appia 20
1211 Geneva 27
Switzerland
Website: http://www.eho.int
E-mail: inf@who.int

World Parrot Trust United States

P.O. Box 50733
Saint Paul, MN 55150
Telephone: (651) 994-2581
Fax: (651) 994-2580
E-mail: usa@worldparrottrust.org

United Kingdom

Karen Allmann, Administrator,
Glanmor HouseHayle,
Cornwall TR27 4HY,
United Kingdom
E-mail: uk@worldparrottrust.org

Australia

Mike Owen
7 Monteray Street
Mooloolaba, Queensland 4557, Australia
E-mail: australia@worldparrottrust.org
Website: http://www.world parrottrust.org

World Resources Institute

1709 New York Avenue NW
Washington, D.C. 20006
Telephone: (202) 638-6300
E-mail: info@wri.org
Website: http://www.wri.org/wri/biodiv

World Society for the Protection of Animals (WSPA)

P.O. Box 190
Jamaica Plain, MA 02130
Website: http://www.wspa.org

United Kingdom Division
Website: http://www.wspa.org.uk/home.html

World Wildlife Fund, US (WWF)

1250 Twenty-fourth Street NW
P.O. Box 97180
Washington, D.C. 20077-7180
Telephone: (800) CALL-WWF
Website: http://www.worldwildlife.org

WorldWatch Institute

1776 Massachusetts Avenue NW
Washington, D.C. 20036
Telephone: (202) 452-1999
Website: http://www.worldwatch.org/
Publications: *WorldWatch, State of the World, Vital Signs* (annuals)

Zero Population Growth

1400 Sixteenth Street NW, Suite 320
Washington, D.C. 20036
Telephone: (202) 332-2200
Fax: (202) 332-2302
E-mail: zpg@igc.apc.org
Website: http://www.zpg.org

Zoe Foundation

983 River Road
Johns Island, SC 29455
Telephone: (803) 559-4790
E-mail: savage@awod.com
Website: http://www.2zoe.com

INDEX

f indicates figures and photos; t indicates tables

1: Earth Systems, Volume I
2: Resources, Volume II
3: People, Volume III
4: Human Impact, Volume IV
5: Sustainable Society, Volume V

1: Earth Systems, Volume I
2: Resources, Volume II
3: People, Volume III
4: Human Impact, Volume IV
5: Sustainable Society, Volume V

1: Earth Systems, Volume I
2: Resources, Volume II
3: People, Volume III
4: Human Impact, Volume IV
5: Sustainable Society, Volume V

1: Earth Systems, Volume I
2: Resources, Volume II
3: People, Volume III
4: Human Impact, Volume IV
5: Sustainable Society, Volume V

1: Earth Systems, Volume I **2:** Resources, Volume II **3:** People, Volume III **4:** Human Impact, Volume IV **5:** Sustainable Society, Volume V

ABOUT THE AUTHORS

JOHN MONGILLO is a noted science writer and educator. He is coauthor of *Encyclopedia of Environmental Science*, and *Environmental Activists*, both available from Greenwood.

PETER MONGILLO has won several awards for his teaching, including School District Teacher of the Year, National Endowment for the Humanities Fellowship Award, and the National Council for Geographic Education Distinguished Teacher Award.